Carbon Nanostructures

More information about this series at http://www.springer.com/series/8633

Rasel Das

Nanohybrid Catalyst based on Carbon Nanotube

A Step-By-Step Guideline from Preparation to Demonstration

Springer

Rasel Das
Leibniz Institute of Surface Modification
Leipzig
Germany

ISSN 2191-3005 ISSN 2191-3013 (electronic)
Carbon Nanostructures
ISBN 978-3-319-58150-7 ISBN 978-3-319-58151-4 (eBook)
DOI 10.1007/978-3-319-58151-4

Library of Congress Control Number: 2017940310

Printed on acid-free paper

This Springer imprint is published by Springer Nature
The registered company is Springer International Publishing AG
The registered company address is: Gewerbestrasse 11, 6330 Cham, Switzerland

I dedicate this book to my father, Milon Kanti Das, and my mother, Fulu Rani Das. Their continual blessings and unflagging support have made possible whatever success I have experienced in my life. I am also much obliged to my elder brother, Rajib Das, who rescued me after I lost my track and gave me a new start when I needed.

Preface

Catalyst is a workhorse for water purification. It works through three main processes, such as photocatalysis, catalytic wet air oxidation and biocatalysis in the field of water treatment. Although a significant number of advances has been noticed in the field of former two processes, biocatalysis has remained opaque in the field of water purification. Enzyme is an effective green biocatalyst and protein in nature which has several quintessential properties, such as multi-substrate specificity, high sensitivity, active in milder reaction conditions, wide pH, and temperature ranges, low reaction times, less by-products formation, and eco-friendly. But free enzymes are not stable and are sensitive to mechanical stresses. In order to improve their stability and reusability, enzymes should be immobilized onto various supports. This book aims to show how to prepare a nanomaterial matrix for enzyme immobilization. This book is intended to give beginning scientists grassroots scientific knowledge from the preparation of NanoBiohybrid (i.e., enzyme immobilized onto carbon nanotube) to the demonstration of water purification. The beauty and intricacy of carbon nanotube purification and functionalization have been presented before developing the NanoBiohybrid catalyst. The book might have an appeal to nanotechnologists, physicists, chemists, environmentalist, biologists, water specialists, chemical engineers, and industrialists who are closely working in the field of catalysis. I hope to help you eat this elephant by giving you one bite at a time.

Leipzig, Germany

Rasel Das

Contents

1 **Introduction** 1
1.1 Background 1
1.2 Why Fresh Water Facility has Shrunk Day by Day? 2
1.3 Can We Tackle the Water Pollution? 3
1.4 Carbon Nanotube: an Enchant Nanomaterial 4
1.4.1 Carbon Nanotube Synthesis 9
1.4.2 Carbon Nanotube Growth Mechanisms 10
1.5 Enzyme: an Unsung Hero for Water Purification 12
1.5.1 Protocatechuate 3,4-dioxygenase: an Enzyme for Water Treatment 13
1.5.2 Protocatechuate 3,4-dioxygenase Immobilized on Carbon Nanotube 13
1.6 What are the Problems that We Fixed Here? 15
1.6.1 Purification of Pristine Carbon Nanotube 15
1.6.2 Solubilization of Carbon Nanotube 16
1.6.3 Development and Characterizations of Nanobiohybrid Catalyst 16
1.6.4 Nanobiohybrid Catalyst for Pollutant Removal 17
1.7 Conclusions 18
References and Future Readings 18

2 **Carbon Nanotube in Water Treatment** 23
2.1 Background 23
2.2 Water Scarcity and Its Consequences 25
2.3 Threats to Conventional Water Treatment Technologies 30
2.4 Carbon Nanotube Scaffold Functions in Water Purifications 31
2.4.1 Adsorption 31
2.4.2 Hybrid Catalysis 33
2.4.3 Desalination 41
2.4.4 Disinfection 42

2.4.5 Sensing and Monitoring 43
2.4.6 Research Gaps and Challenges 44
2.5 Conclusions 46
References and Future Readings 47

3 Carbon Nanotube Purification 55
3.1 Background 55
3.2 Literature Review 56
3.2.1 What are the Impurities Commonly Associated with the Pristine CNT? 56
3.2.2 What Kinds of Characterization Methods are Commonly Used for CNT Impurities Detection?. 57
3.2.3 What are the Methods Necessary for CNT Purification? 57
3.2.4 What Challenges Does Exist with CNT Purification? 58
3.3 Materials and Methods 60
3.3.1 Materials and Reagents 60
3.3.2 Instrumentation 60
3.3.3 Wet Chemical Treatments of MWCNT. 60
3.4 Results 61
3.4.1 Predicted Chemical Reactions of HCl, H_2O_2, and KOH with MWCNT. 61
3.4.2 TEM Analysis. 63
3.4.3 EDX Analysis. 64
3.4.4 ATR-IR Analysis 65
3.4.5 Raman Spectroscopy Analysis 66
3.4.6 TGA Analysis. 66
3.4.7 Colloidal Stability Measurement 68
3.5 Discussion 68
3.6 Conclusions 70
References and Future Readings 71

4 Carbon Nanotube Functionalizations. 75
4.1 Background 75
4.2 Literature Review 76
4.2.1 Is CNT Hydrophobic in Nature?. 76
4.2.2 What are the Strategies Commonly Used for CNT Solubilization? 77
4.2.3 Non-covalent Modification of CNT. 79
4.3 Materials and Methods 82
4.3.1 Materials and Reagents 82
4.3.2 Instrumentation 83
4.3.3 Wet Chemical Oxidations of MWCNT 83

4.4 Results ... 84
4.4.1 Predicted Chemical Reactions of HNO_3/H_2O_2 and $KMnO_4$ with MWCNT ... 84
4.4.2 XPS Analysis of MWCNT ... 86
4.4.3 Titration Analysis ... 87
4.4.4 TEM Analysis ... 90
4.4.5 Raman Spectroscopy Analysis ... 91
4.4.6 TGA Analysis ... 92
4.4.7 Colloidal Stability of MWCNT ... 93
4.5 Discussion ... 95
4.6 Conclusions ... 100
References and Future Readings ... 100

5 Nanobiohybrid Preparation ... 105
5.1 Background ... 105
5.2 Literature Review ... 107
5.2.1 Enzyme Immobilization onto CNT Using Non-covalent Approach ... 108
5.2.2 Enzyme Immobilization onto CNT Using Covalent Approach ... 111
5.2.3 Target Enzyme for Attachment with the CNT ... 112
5.3 Materials and Methods ... 113
5.3.1 Materials and Reagents ... 113
5.3.2 Instrumentation ... 114
5.3.3 Preparation of F-MWCNT ... 114
5.3.4 Preparation of Nanobiohybrid ... 115
5.4 Results ... 116
5.4.1 Characterization of F-MWCNT ... 116
5.4.2 Characterization of Nanobiohybrid ... 117
5.4.3 Optimization of Nanobiohybrid ... 120
5.5 Discussion ... 121
5.6 Conclusions ... 124
References and Future Readings ... 124

6 Nanobiohybrid for Water Treatment ... 129
6.1 Background ... 129
6.2 Literature Review ... 130
6.3 Materials and Methods ... 133
6.3.1 Materials and Reagents ... 133
6.3.2 Free 3,4-POD and Nanobiohybrid Activity Assays ... 133
6.3.3 Free 3,4-POD and Nanobiohybrid Stability Assays ... 134
6.3.4 Removal of 3,4-DHBA ... 134
6.3.5 Data Analysis ... 135

6.4 Results 135
6.4.1 Effects of pH and Temperature on Free 3,4-POD and Nanobiohybrid Activities 135
6.4.2 Kinetic Analyses of Free 3,4-POD and Nanobiohybrid 136
6.4.3 Stability Studies of Free 3,4-POD and Nanobiohybrid 137
6.4.4 Degradation Kinetics of 3,4-DHBA 137
6.5 Discussion 139
6.6 Conclusions 141
References and Future Readings 141

Abbreviations

AFM	Atomic Force Microscopy
ATR-IR	Attenuated Total Reflectance Infrared
ABTS	2,2′-Azino-Bis-(3-Ethylbenzthiazoline-6-Sulfonic Acid)
AC	Activated Carbon
ACS	American Chemical Society
AD	Arc Discharge
α	Alpha
Al	Aluminum
at	Atomic
$(NH_4)_2S_2O_8$	Ammonium Persulfate
NH_3	Ammonia
NH_4OH	Ammonium Hydroxide
Ar	Argon
&	And
BCA	Bicinchonic Acid
BSA	Bovine Serum Albumin
B. subtilis	*Bacillus subtilis*
β	Beta
CNT	Carbon Nanotube
Ch	Chiral
C	Carbon
CVD	Chemical Vapor Deposition
CWAO	Catalytic Wet Air Oxidation
Con A	Concanavalin A
CNTECE-Tyr	Carbon Nanotube-Epoxy Composite Electrode-Tyrosinase
CHES	2-(Cyclohexylamino) Ethanesulfonic
CABS	4-(Cyclohexylamino)-1-Butanesulfonic
CB	Conduction Band
COOH	Carboxylic
cm	Centimeter

cm^2	Centimeter Square
C_{60}	Carbon 60
Cu	Copper
Co	Cobalt
CO_2	Carbon Dioxide
CD	Circular Dichroism
C=O	Carbonyl
DHBA	3,4-Dihydroxybenzoic Acid
DTA	Differential Thermal Analysis
d	Diameter
DOA	Dissociated Oxygen Atom
DM	*n*-Dodecyl-β-D-Maltoside
DBT	Dibenzothiophene
D	Defective
N_2O_6	Dinitrogen Hexaoxide
ΔE_g	Delta Energy
°C	Degree Celsius
EDX	Energy-Dispersive X-ray Spectroscopy
EC	Enzyme Commission
E. coli	*Escherichia coli*
eq	Equation
e^-	Electron
e*V*	Electron Volt
E_0	Energy
EDC	1-ethyl-3-(3-dimethylaminopropyl)carbodiimide
e.g.	Exemplī Grātiā
etc.	Etcetera
et al.	Et alia
ε	Epsilon
COC	Epoxy
FTIR	Fourier Transform Infrared
FWCNT	Few-walled Carbon Nanotube
FE-SEM	Field Emission Scanning Electron Microscope
F-MWCNT	Functionalized Multi-walled Carbon Nanotube
GDP	Gross Domestic Product
GECE-Tyr	Graphite Epoxy-Composite Electrode Tyrosinase
G	Graphite
g	Gram
>	Greater Than
≥	Greater Than Equal
HRP	Horseradish Peroxidase
HRTEM	High-Resolution Transmission Electron Microscope
HQD	Hydroxyquinol 1,2 Dioxygenase
HEPES	4-(2-hydroxyethyl)-1-piperazineethanesulfonic Acid
HCl	Hydrochloric Acid

H_2O_2	Hydrogen Peroxide
H_2SO_4	Sulfuric Acid
HNO_3	Nitric Acid
H	Hydrogen
h	Hour
h^+	Hole
OH	Hydroxyl
$^{\bullet}OH$	Hydroxyl Radical
IWMI	International Water Management Institute
ID	Identification
IR	Infrared
I	Intensity
i.e.	id est
Fe	Iron
KHz	Kilohertz
kDa	Kilo Dalton
kV	Kilovolt
Kg	Kilogram
Kev	Kilo Electron Volt
LA	Laser Ablation
LBL	Layer-By-Layer
L	Liter
<	Less Than
MDG	Millennium Development Goals
MF	Micro Filtration
MWCNT	Multi-walled Carbon Nanotube
MMCNT	Mixed Matrix Carbon Nanotube
MOX	Methyl Paraoxon
MD	Microbial Decontamination
MES	2-(*N*-morpholino)ethanesulfonic Acid
ml	Milliliter
μl	Microliter
M	Molar
mM	Millimolar
μM	Micromolar
mmole	Millimole
μmole	Micromole
MΩ	Microohm
mg	Milligram
μg	Microgram
min	Minute
m^2	Meter Square
m^3	Meter Cube
mm^2	Millimeter Square
Mpa	Megapascals

μm	Micrometer
mA	Milliampere
Mn	Manganese
Mg	Magnesium
NP	Nanoparticle
NHS	*N*-Hydroxysuccinamide
NF	Nanofiltration
Ni	Nickel
NO_3^-	Nitrate
NO_2	Nitrogen Dioxide
N	Nitrogen
nmol	Nanomole
nm	Nanometer
O-MWCNT	Oxidized Multi-walled Carbon Nanotube
O	Oxygen
OP	Organophosphate
O_3	Ozone
O_2	Oxygen Molecule
$O_2^{\bullet}$	Oxygen Radical
O-acylisorea	Ortho acylisorea
P. putida	*Pseudomonas putida*
POP	Persistent Organic Pollutants
ppb	Parts Per Billion
p-Cresol	Para Cresol
P. aeruginosa	*Pseudomonas aeruginosa*
POU	Point of Use
POG	Point of Generation
PDB	Protein Data Bank
Pd	Palladium
Pt	Platinum
PEG	Poly (Ethylene Glycol)
%	Percentage
H^+	Proton
MnO_4^-	Permanganate
pH	*pi:heɪt*
π	pi
ψ	Psi
KBr	Potassium Bromide
KOH	Potassium Hydroxide
$KMnO_4$	Potassium Permanganate
RO	Reverse Osmosis
r	Radius
ROS	Reactive Oxygen Surface
RE	Removal Efficiency
RT	Room Temperature

rpm	Revolution Per Minute
REFS	References
Ru	Ruthenium
SWCNT	Single-walled Carbon Nanotube
SEM	Scanning Electron Microscope
SAR	Special Administrative Region
STM	Scanning Tunneling Microscope
S. aureus	*Staphylococcus aureus*
sp.	Species
SBP	Soybean Peroxidase
Sdn Bhd	Sendirian Berhad
NaOH	Sodium Hydroxide
NaCl	Sodium Chloride
Na^{2+}	Sodium
Ag	Silver
TEM	Transmission Electron Microscope
T	Translation
TGA	Thermal Gravimetric Analysis
TiO_2	Titanium Dioxide
Θ	Theta
2,4-DCP	2,4-Dichloro Phenol
2,4,6-TCP	2,4,6-Trichloro Phenol
3,4-DHBA	3,4-Dihydroxybenzoic Acid
3,4-POD	Protocatechuate 3,4-dioxygenase
2D	Two Dimensions
3D	Three Dimensions
UF	Ultra Filtration
UNICEF	United Nations Children's Fund
UV-vis	Ultraviolet–visible
UV	Ultraviolet
U	Unit
UN	United Nations
UNDP	United Nations Development Programme
USA	United States of America
US	United States
UK	United Kingdom
v/v	Volume/Volume
vol	Volume
V	Voltage
VB	Valence Band
VACNT	Vertically Aligned Carbon Nanotube
V_{max}	Velocity Maximum
WHO	World Health Organization
WWAP	World Water Assessment Programme
wt	Weight

H_2O	Water
XPS	X-ray Photoelectron Spectroscope
XRD	X-ray Powder Diffraction
ZnO	Zinc Oxide

Chapter 1
Introduction

Live as if you were to die tomorrow. Learn as if you were to live forever.
—Source: Mahatma Gandhi: *Autobiography: The Story of My Experiments with Truth* by Mohandas Karamchand Gandhi, 1983.

Abstract The efficient handling of both the persisting and newly emerging pollutants is a must, since they are continuously defiling the limited fresh water resources, seriously affecting the terrestrial, aquatic, and aerial flora and fauna. The pressing need to overcome current major limitations of advanced oxidation processes (AOP), such as energy-intensive, toxic intermediates production, less selectivity and sensitivity for dilute solutions and catalyst leaching effects have motivated us to establish a different route for water purification called "NanoBiohybrid Catalyst" technology. Although enzymes have been used for a long time to treat wastewater, they are not stable, have low life span, highly sensitive to mechanical stresses and difficult to separate from the substrates. In order to overcome these drawbacks, this book shows how to use carbon nanotube (CNT) as an excellent support matrix for enzyme immobilization. Unfortunately, raw CNT are hydrophobic and often contaminated with various impurities, such as amorphous carbons, metals and ashes which hinder its conjugation with enzymes. Thence this book displays first how to use simple chemical methods for CNT purification and also functionalization with bioconjugating functionalities for water dispersion properties. The book then reveals the methods based on which one can immobilize enzymes onto the purified and functionalized CNT to birth a NanoBiohybrid Catalyst. Finally, the potentiality of the hybrid catalyst for organic pollutants removal from the water has been demonstrated.

1.1 Background

The rhyme "Water, water, everywhere, nor any drop to drink" by Samuel Taylor Coleridge depicted fresh and safe water crises in the middle of huge and vast sea water in 1798 [16]. Unfortunately, the problem of water scarcity has been sustained

R. Das, *Nanohybrid Catalyst based on Carbon Nanotube*,
Carbon Nanostructures, DOI 10.1007/978-3-319-58151-4_1

currently, since Ban Ki-moon (Secretary-General of the United Nations) has implied water shortages of the twenty-first century that could weaken the Millennium Development Goals (MDG) [80], indicating exacerbation of this ubiquitous problem in the near future. Moreover, world population has been experienced to increase at a rate of 80 million/year, putting an additional demand of 64 billion cubic meters of potable water per annum [92, 93]. In 2013, it has been observed that about 768 million people in the world suffered from improved fresh water facilities [89] of which Asia's populace counts 380 million [83]. The United Nations (UN) has claimed that almost 2.0 billion people could not use fresh, clean and safe water in 2013 [81]. It has been forecasted that by 2025 about 1.8 billion people will be forced to survive under absolute water scarcity [82]. On the other hand, the sea level can rise by about 10–15 inches by the 2025, and salt water would obtrude into the estuary ultimately increasing salinity [58]. These paint a scary picture of water scarcity, which is expensive, since it jeopardizes the industrial production, and diminishes the availability of hygienic foods, drinks, and indirectly causes various epidemic diseases such as dengue, malaria, hepatitis, etc. Deficiency of clean and safe water counted about 3.1% deaths worldwide [87], which is forecasted to rise over the years. According to World Health Organization (WHO) and the United Nations Children's Fund (UNICEF), one child has been died in every 15 s due to the direct or indirect consumption of unsafe water [88]. Thus an effective water purification technology is the need of the day, and its failure might further endanger the life processes and eco-friendly human existence.

1.2 Why Fresh Water Facility has Shrunk Day by Day?

Industrialization, deforestation, and urbanization have consistently been polluted finite fresh water resources by adding persistent pollutants to world water bodies through point and nonpoint origins; leading to a global decrease in the availability of clean and safe water. The pollutants including gas fumes (carbon dioxide, sulfur dioxide, nitric oxide, and so on); inorganic species (arsenic, cadmium, chromium, cobalt, copper, selenium, and other miscellaneous metals); organic entities (various dyes, fertilizers, pesticides, and hydrocarbons); and biological contaminants (viruses, bacteria, and fungi) have been constantly dwindling the availability of clean water for terrestrial, aquatic, and aerial, flora and fauna. Among these waste water contaminants, phenolic water pollutants are notorious because of its wide availability, persistence, and bactericidal effects [30, 48]. The pollutants have been appended rapidly with the conventional priorities from both the natural and anthropogenic sources. For instance, the 3,4-Dihydroxybenzoic acid (3,4-DHBA), simply known as Protocatechuic acid—is an important toxic phenolic water pollutant that has been largely found in foods processing waste water effluents [66], especially in olive oil factory [62] and also secreted by various plant species [6] into water. It has been noticed that the concentration of 3,4-DHBA is calculated as polyphenolic group, which was 4–10 mg/L especially in olive oil mill waste water [35]. Therefore, such phenolic mixtures could compete with each other in the degradation process and the accurate degradation rate would be impaired in complex matrix [29]. The situation has become worst from hours-to-day, since people are continuously damping their wastes and also comes from the decompositions of flora and fauna. As a corollary, they are known to be present at high concentration in various waste water matrices [66]. The 3,4-DHBA has shown contradictory biological effects on animal and human cell lines. On the one hand, it has shown oxidative stress, tumor production, hepatotoxicity, neurotoxicity, and inflammation in mice and human tissues [6, 53, 54]. On the contrary, it could inhibit chemical-actuated carcinogenesis of different mouse tissues such as liver, kidney, pancreas, and so on [79]. However, such effects have not observed in human cells [6]. Therefore, effective and inexpensive regulatory tool should be developed to control the quantity of 3,4-DHBA in public sewage systems.

1.3 Can We Tackle the Water Pollution?

In order to tackle wide range of pollutants, different water treatment technologies have been developed and applied at both experimental and field levels. The technologies are commonly classified as primary (screening, filtration, centrifugal, separation, sedimentation, coagulation, flocculation, etc.); secondary (aerobic and anaerobic treatments) and tertiary (distillation, crystallization, evaporation, solvent extraction, oxidation, precipitation, ion exchange, microfiltration (MF), ultrafiltration (UF), nanofiltration (NF), reverse osmosis (RO), adsorption, electrolysis, electrodialysis, and so forth) [33]. Out of these, only a few are capable of solving the water treatment issues. Some methods have found energy and operationally intensive and cannot be affordable at commercial level [19]. Adsorption technique is easy and simple, yet its ability for water purification is not satisfactory [96].

The advent of nanotechnology has given immeasurable opportunities to purify water even at ionic state. The various nanostructured materials such as fullerenes, nanofibers, nanowires, zeolites, and various nanoparticles have been fabricated with features such as high aspect ratio, reactivity, tunable pore volume, electrostatic, and hydrophilic and hydrophobic interactions which are useful in adsorbing, catalyzing, sensing and mitigating of different water pollutants. In detail, titanium dioxide nanoparticles have been used in micropollutants transforming redox reactions [44]. Photocatalytic nano-TiO_2 could detoxify phenolic recalcitrant compounds, microbial and odorous chemicals into harmless species [14, 67]. Other nanoparticles such as silver, gold, iron, and their oxides have been widely used in water purifications. Gold and iron nanoparticles have been used for removing inorganic heavy metals from waste waters [45]. Most of the nanomaterials have increased surface porosity which improves salt rejection and avoids macrovoid formation [31]. For example, the CNTs have been used to make composites of RO, UF, NF, and MF membranes. This enhances salt retention ability, curtails costs, land area, and energy for brackish and sea water desalinations [19]. Zeolite nanoparticles have been mixed with polymer matrix to form a thin film RO membrane [23]. It has been increased water transports and retained >99.7% of salts [42]. Silica nanoparticles have been doped into RO polymer matrices for water desalination [41]. It improves polymer's networks, pore diameters, and transport properties. Silver nanoparticles have shown good properties for disinfecting biological pollutants such as bacteria, viruses, and fungi [45].

Although several nanomaterials have been appreciated in water purification [45], they are not free from limitations. Drawbacks include thermal instability, high pressure requirements, fouling, pollutant precipitation, pore blocking, low water influx, slow reaction rate, toxic by-products, and aggregation on storage [17]. Low chances of reusability and unknown risks to eco-systems have remained major concerns [45]. Several hurdles such as insufficient data on long time operation, stability and scaling up have to be resolved prior to commercialization [31]. In order to overcome current shortcomings, nanomaterial needs to be fabricated and manipulated for increasing their surface reactivity, selectivity, and sensitivity toward wide range of water contaminants. Therefore, effective fabrication and functionalization strategies are always appreciable to use nanomaterials in a unit operation process of water purification fields.

1.4 Carbon Nanotube: an Enchant Nanomaterial

In group IV of the periodic table, Carbon (C) is the first element which has a $1s^2 2s^2 2p^2$ electronic configuration. The orbital shows that four valence electrons that might allow it to form a number of hybridized atomic orbitals. C has the capability to yield various nanostrctures because they can share electron pairs and form covalent bonds. The bonds have directional properties; this in turn provides C the capability to form various solid structures. Hence the physicochemical properties of C-structures depend on the nature of the covalent bonds that are formed. CNT is a C-based nanomaterial and is a more exotic entity in the field of nanotechnology has proven them important materials for various applications. For example, CNT has been extremely used in waste water treatment plants and their suitability for handling organic, inorganic, and biological water pollutants is commendable [84]. The tubes have effectively been used in adsorption [84], catalysis, separation, disinfection [17], and sensor [3] technologies due to its large surface area, high aspect ratio, chemical reactivity and so on for detecting and mitigating of different water pollutants. The CNT is composed of graphitic sheet (allotropic form of carbon) rolled up in a cylindrical shape tube like structure with the appearance of latticework fence [39] (Fig. 1.1). Single-walled carbon nanotube (SWCNT) consists of a single graphene shell [39, 40] (Fig. 1.1a), whereas multi-walled carbon nanotube (MWCNT) is composed of multiple layers of graphene sheets (Fig. 1.1b).

Figure 1.2 shows the formation of a SWCNT by rolling up a single graphene sheet. In SWCNTs the C-bonding that exists is similar to that which exists in graphite as opposed to that found in diamond. For example, diamond has a coordination number of four, with sp^3 hybridization, while, on the other hand, sp^2 hybridization exists in the planar layers of C atoms that give graphite its structure,

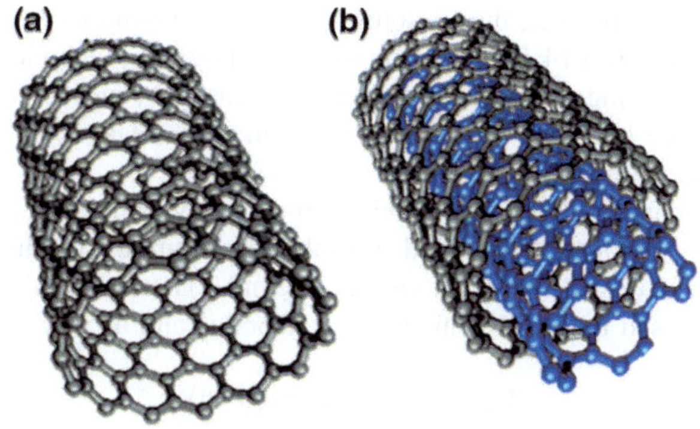

Fig. 1.1 (Super) structure representations of **a** a SWCNT and **b** a MWCNT. Figure is adapted with permission from American Chemical Society (ACS) [102]

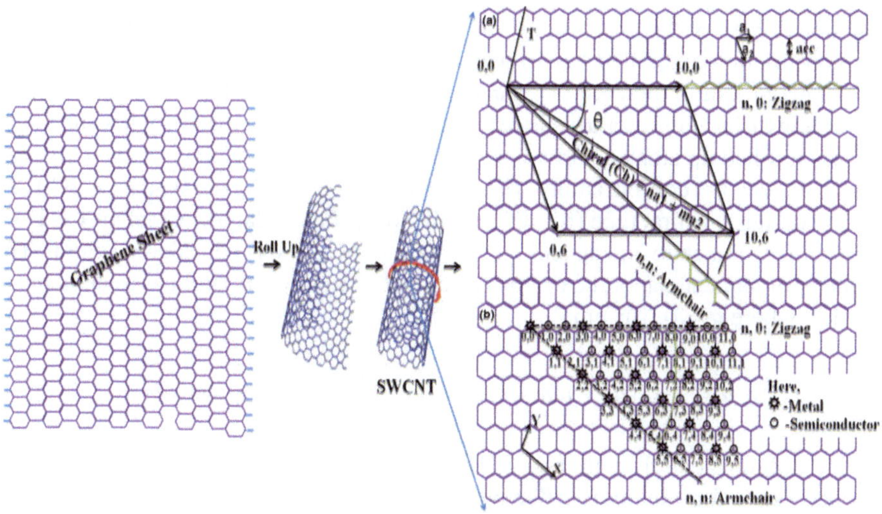

Fig. 1.2 Schematic construction of SWCNT from a graphene sheet. **a** C_h, depicts chiral vector; and a_1 and a_2, represent the unit cell vectors of the hexagonal cell. The chiral angle (θ) is the angle between zigzag axis and chiral vector (C_h) and could be represented by $\theta = \tan^{-1}$ $(m\sqrt{3}/m+2n)$; **b** Representation of the integer values (*n, m*) and paired of them decides zigzag, armchair and chiral SWCNTs

and in the bonding that leads to the tubular structure of SWCNTs. Each carbon atom contributes 3 electrons to 3 equivalent sigma bonds within the plane and has 1 electron left in the perpendicular pz orbitals; such electrons are delocalized over the entire plane, resulting in a π-electron orbital system that allows the fourth valence electron to essentially move freely over the plane. Individual nanotubes naturally align themselves into "ropes" held together by van der Waals forces, more specifically, π -stacking. Within the layers, the C–C bond distance is similar to the bond length in benzene (i.e., the C atoms are strongly bound to each other and the C–C distance is about 0.14 nm), leading to a very large in-plane value for Young's modulus. However, the distance between layers (ca., 0.34 nm) is sufficiently large that the layers are bounded to each other mainly by weak, long range Van der Waals type interaction.

One dimensional SWCNT commonly expressed by chiral (C_h) and translation (T) vectors. The vector has sharpened from the first atom toward the second one in 2D graphene sheet (Fig. 1.2a). Both vectors (C_h and T) determine CNT unit size and commonly expressed by the Bravais lattice vectors $\mathbf{a}_1$ and $\mathbf{a}_2$ as Eqs. 1.1 and 1.2.

$$\mathrm{C_h} = n\mathrm{a}_1 + m\mathrm{a}_2 = (n, m) \tag{1.1}$$

$$\mathrm{T} = t_1\mathrm{a}_1 + t_2\mathrm{a}_2 = \frac{n + 2m, -2n - m}{W}, \tag{1.2}$$

where a_1 and a_2 are the unit vectors; n and m are the integers, and W is the greatest common divisor of the expressions $n + 2m$ and $2n + m$. Those nanotubes following this construction are called (n, m) nanotubes (i.e., SWCNTs).

The magnitude of the vector $\mathbf{C_h}$ can be used to calculate CNT radius by $R = C_h/2\pi$. According to the integer values, three types of CNT could be appeared when the sheets of carbon rolled up along a particular single symmetry axis. When $n = m$, it takes an Armchair ($\theta = 0°$) conformation; when $m = 0$, it becomes Zigzag type ($\theta = 30°$) and for $n \neq m$, it is a Chiral tube ($\theta = 0°$ and $30°$) [65] (Fig. 1.2a, b). Armchair CNT is a mirror image match to zigzag and vice versa and therefore is considered achiral. The diameter of an ideal nanotube can be calculated from its (n, m) indices as depicted in Eq. 1.3, where $a = 0.246$ nm.

$$d = \frac{a}{\pi}\sqrt{(n^2 + nm + m^2)} = 78.3\sqrt{((n + m)^2 - nm)pm} \tag{1.3}$$

The different chiral angle and CNT diameters bring different properties which are necessary for the more effective and emerging applications of CNTs. The symmetry operation of CNT structures can be expressed as $R = (\Psi/T)$, and the number of hexagons can be expressed as N and is determined by the value of n, m according to the Eq. 1.4.

$$N = 2\left(m^2 + n^2 + nm\right)/\mathrm{dR}, \tag{1.4}$$

where dR is CNT diameter. Each hexagon consists of two carbon atoms (Fig. 1.2a). The metallic and semiconducting CNTs defined by the integers (n, m) as shown in Fig. 1.2b. If the chiral indices are equal, $n = m$, the CNT is metallic; if $n - m$ is a multiple of 3, then the CNT is semiconducting, with a very small band gap; otherwise, the CNT is a moderate semiconductor. Interestingly, some CNTs have conductivities higher than that of copper, while others behave more like silicon.

Double-walled carbon nanotubes (DWCNTs) form a special class of nanotubes because their morphology and properties are similar to those of SWCNTs but they are more resistant to chemicals. This is especially important when it is necessary to graft chemical functions to the surface of the nanotubes (functionalization) to add properties to the CNT. Covalent functionalization of SWCNTs will break some C=C double bonds, leaving "holes" in the structure on the nanotube, and thus modifying both its mechanical and electrical properties. In the case of DWCNTs, only the outer wall is modified.

There are two models that can be used to describe the structures of MWCNT. In the *Russian Doll* model, sheets of graphite are arranged in concentric cylinders, e.g., a (0, 8) SWCNT within a larger (0, 17) SWCNT. In the *Parchment* model, a single sheet of graphite is rolled in around itself, resembling a scroll of parchment or a rolled newspaper. The interlayer distance in MWCNT is close to the distance between graphene layers in graphite, approximately 3.4 Å. The Russian Doll structure is observed more commonly. Its individual shells can be described as SWCNTs, which can be metallic or semiconducting. Because of statistical probability and restrictions on the relative diameters of the individual tubes, one of the shells, and thus the whole MWCNT, is usually a zero-gap metal. Since MWCNTs are constructed by joining SWCNTs, so the models for SWCNTs can also be used to describe MWCNT. According to "Russian Doll" model, MWCNT looks like coaxial cylinder curve, coaxial polygonized or scroll multilayer nanotubes depending on interlayer's structural folding as illustrated in Fig. 1.3 [10, 65, 97]. The model is the most common and popular form of MWCNTs, where the chiral vector and the distance round of the CNT can be expressed by the following Eq. 1.5.

$$C = \mathrm{IC_hI\,a}\sqrt{(n^2 + nm + m^2}, \tag{1.5}$$

where a—is the length of the unit cell vector a_1 or a_2. This length is connected to the (C)–(C) bond length a_{cc} and is determined by the Eq. 1.6.

$$\mathrm{a} = \mathrm{Ia_1I} = \mathrm{Ia_2I} = \mathrm{a_{cc}}\sqrt{3}. \tag{1.6}$$

For graphite, a_{cc}; is 0.1421 nm, *C*, is the length of a chiral vector C_h. By using *C* values, the CNT diameters (D) can be calculated according to Eq. 1.7.

$$\mathrm{D} = \mathrm{C}/\pi. \tag{1.7}$$

The tiny structure of CNT has remained mysterious maze to many scientists because of its unusual complex C-networks. Therefore, developing novel characterization techniques with improved simulation studies could be a way to find out the directions and mechanisms through which their complex carbon framework could build.

A comparative view of different physicochemical properties of SWCNT and MWCNT is revealed in Table 1.1. According to Table 1.1, CNT is the strongest and stiffest materials yet discovered in terms of tensile strength and elastic modulus, respectively. This strength results from the covalent sp^2 bonds formed between the individual C atoms. Although the strength of individual CNT shells is extremely high, weak shear interactions between adjacent shells and tubes lead to significant

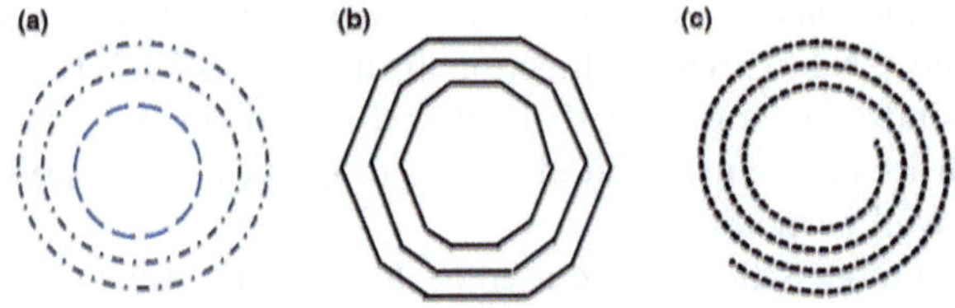

Fig. 1.3 Top views of MWCNTs: **a** cylindrically curved Russian Doll model; **b** coaxial polygon and **c** graphene sheets scroll; designed according to [97]

reduction in the effective strength of MWCNT and CNT bundles down to only a few GPa [24]. CNTs are not nearly as strong under compression. Because of their hollow structure and high aspect ratio, they tend to undergo buckling when placed under compressive, torsional, or bending stress. The major conclusions that we can draw here are owing to the material's exceptional strength and stiffness, CNTs have

Table 1.1 Comparison of physicochemical properties of SWCNT and MWCNT

Features	SWCNT	MWCNT
Structure type	Single wall	Multi wall
Graphene layer	Single	Multiple
Diameter	1 to $\sim$2 nm	$\sim$2 to $\geq$50 nm
Elasticity	$\sim$1 TPa	$\sim$0.2–0.95 TPa
Tensile strength	50–500GPa	11–63 GPa
Specific surface area	$\sim$400–900 m^2/g	$\sim$200–400 m^2/g
Thermal conductivity	3000 W m^{-1} K^{-1}	3000 W m^{-1} K^{-1}
Thermal stability (air)	>700 °C	>700 °C
Resistivity	10^{-4}–10^{-3}[Ω m]	1.8×10^{-5}–6.1×10^{-5} [Ω m]
Specific gravity	>1 g/cm^3	0.8 g/cm^3–1.8 g/cm^3
Growth mechanism	Open/close ended lip–lip interactions	Catalytic growth action
Folding	Easy	Hard
Catalyst required	Yes	No
Characterization	Relatively easy	Complex structure
Amount	Difficult to get many	Relatively easy to get
Functionilization	May produce defect sites	Less chance to produce defect sites
Purity	Low	High
Dispensability	Low	High
Body accumulation	Poor	High
Major application	Spintronics and optoelectronics	Adhesive and Coatings

been constructed with length-to-diameter ratio of up to 132,000,000:1, significantly larger than for any other material. In addition, owing to their extraordinary thermal conductivity, mechanical, and electrical properties; CNTs find applications as additives to various structural materials.

Standard SWCNT can withstand a pressure up to 25 GPa without deformation. They then undergo a transformation to superhard phase nanotubes. Maximum pressures measured using current experimental techniques are around 55 GPa [24]. However, these new superhard phase nanotubes collapse at an even higher pressure. The bulk modulus of superhard phase nanotubes is 462–546 GPa, even higher than that of diamond (420 GPa for single diamond crystal).

The surface wettability of CNT is of importance for its applications in Nanobiohybrid preparation. Although the intrinsic contact angle of graphite is around 90°, the contact angles of most as-synthesized CNT arrays are over 160°, exhibiting a superhydrophobic property. If a system used to eliminate pollutants out of water, one might bring water and/or pollutants to the reactive surfaces of the Nanobiohybrid catalyst. By applying a voltage as low as 1.3 V, the extreme water repellant surface can be switched to a superhydrophilic one [94].

As we discussed above unlike graphene, which is a two-dimensional semimetal, CNT are either metallic or semiconducting along the tubular axis. All armchair ($n = m$) nanotubes are metallic, and nanotubes (4, 6), (1, 9), etc., are semiconducting. The rule regarding metallic versus semiconductor behavior has exceptions, because curvature effects in small diameter tubes can strongly influence electrical properties. Thus, a (5, 0) SWCNT that should be semiconducting in fact is metallic according to the calculations. In theory, metallic CNTs can carry an electric current density of 4×10^9 A/cm^2, which is more than 1000 times greater than those of metals such as copper [38]. CNTs are thus being explored as conductivity enhancing components in photocatalytic water treatment materials and many groups are attempting to expedite the photocatalyst activity for fastest wastewater pollutants degradation. But there are significant challenges to be overcome, however, such as the much more resistive nanotube-to-nanotube junctions and impurities, all of which lower the electrical conductivity of the macroscopic CNTs by orders of magnitude, as compared to the conductivity of the individual CNT.

1.4.1 Carbon Nanotube Synthesis

Figure 1.4 illustrates currently used CNT synthesis methods. Herein among six methods, three of its such as arc discharge (AD), laser ablation (LA), and chemical vapor deposition (CVD) have been commonly used for CNT productions [63]. Since, scientists have been trying to produce CNTs in cheaper ways; many new techniques are coming out every day. Interestingly, most of them have used common experimental parameters such as carbon feedstock, metal catalysts and temperatures. Some methods such as LA and template/bottom-up have been dedicated to the synthesis of SWCNTs, whereas others such as CVD, electrolysis, and

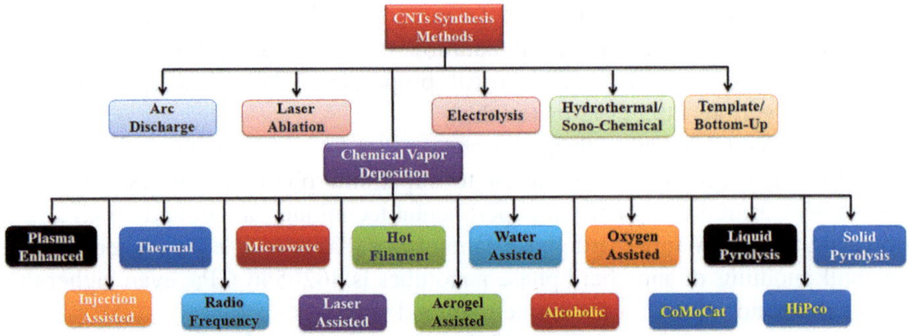

Fig. 1.4 Currently available CNT synthesis methods

hydrothermal methods are more suitable for MWCNT formations. AD is an old method and has been applied for the synthesis of SWCNT and MWCNT. Besides, some interesting literatures have demonstrated the use of rapid and cheap domestic microwave oven techniques such as metal complex mixture [77], nano fiber [77] and nickel nanoparticle methods [59] for producing MWCNTs.

Commercial CNT potential can be estimated from its market production value of US $37 million (SWCNT) and US $630 million (MWCNT) and is expected to reach US $1.1 billion by 2016. Nearly 750 metric tons of CNT were utilized over the past few years for energy and environmental applications and continues to increase. Current prices of SWCNT and MWCNT are US $25–300 and US $0.10–25/g, respectively. While MWCNT may not have some of the superior SWCNT individual properties, the significant cost reduction would more than compensate for performance reduction while still providing an enhanced material. It has been calculated that average costs of SWCNT is US $1.26 and 2.2/g for organic and inorganic pollutants removal, respectively. In comparison, MWCNT will decrease cost to US $0.22 and 0.57/g for organic and inorganic pollutant removal, respectively.

1.4.2 Carbon Nanotube Growth Mechanisms

The AD, LA, and CVD require similar catalysts to grow CNTs. Therefore, they might share a common growth mechanism. The actual growth mechanism is unclear and it has been remained a debatable issue among the scientists. Typically, two general routes, namely base growth [7] (Fig. 1.5a) and tip growth models [8] (Fig. 1.5b) have been documented over the years. The overall process involves three main steps: (i) carbon feedstock is supplied on catalysts surface to get fullerenes as intermediate; (ii) scoot (small carbon fragments like C2 and C3) is generated from the decomposition of hydrocarbons by heat and subsequently is deposited on catalysts surface and (iii) finally nanotube grows from the deposited meta stable carbide particles until the catalysts and carbon fragments are saturated.

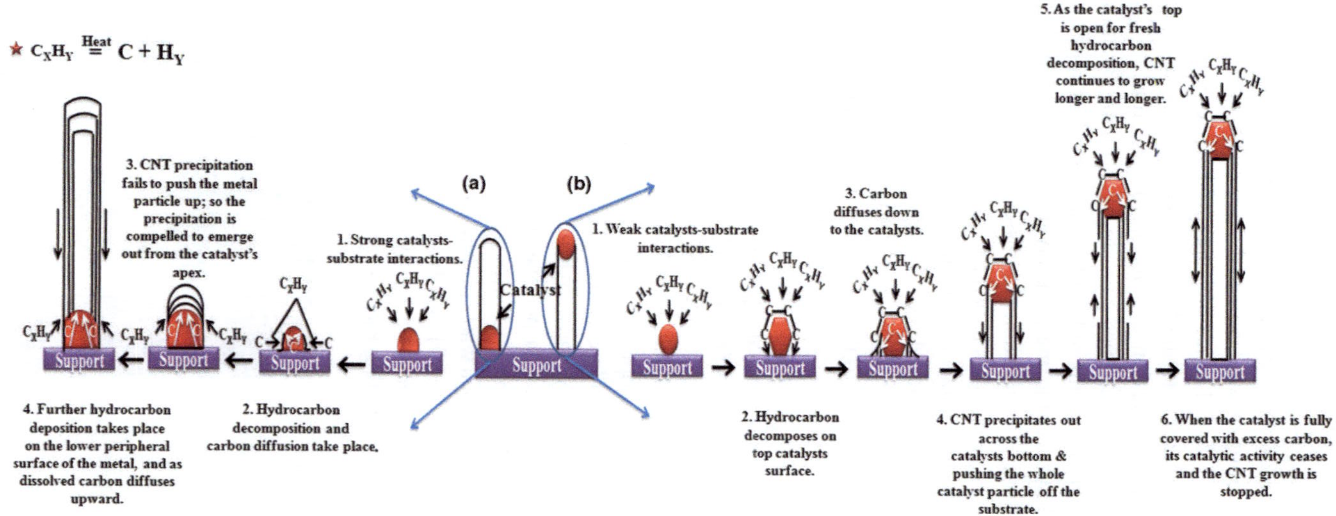

Fig. 1.5 (a) Base and (b) Tip growth mechanisms for CNT formation

1.5 Enzyme: an Unsung Hero for Water Purification

With the appearance of complex waste water pollutants, high cost of specialized chemicals and generation of solid chemical wastes have brought enzyme biocatalysts in water purifications [45, 95], since they are highly specific and efficient [55] to their pollutant substrates. Enzyme is an effective green catalyst and protein in nature that is even active in milder reaction conditions. It has broad substrate, pH and temperature ranges, low reaction times with no byproducts, functioning in harsh conditions and eco-friendly compared with other physical and chemical treatments [2, 45, 70, 75]. However, the main disadvantages of using free enzyme are poor stability and sensitivity to mechanical stresses [12, 25]. In order to improve their stability and reusability, enzymes have been immobilized onto various supports [25] as shown in Fig. 1.6. For instances, enzymes immobilized on mesoporous activated carbon (MAC), graphene and MWCNT have shown that the maximum enzyme loading capacity of lipase was 3.69, 8.63 and 145 μg/mg of support, respectively [64, 78, 100]. This has shown to increase enzyme's effective loading in the order of MWCNT > graphene > AC which have good catalytic activity, higher mechanical stability, broad pH and temperature ranges compared with free enzyme [4, 12, 25]. In addition, recovery of enzyme catalyst is necessary for decreasing its costs for multiple time uses, which could be easier upon immobilization and often impossible with the free state [75]. So, the method is stable and economically viable for developing a novel water purification technology.

Many works have been adopted to immobilize enzymes on CNTs such as tyrosinase [75], horseradish peroxidase (HRP) [50], laccase [52], glucose oxidase [32], and so on [25]. Unfortunately, these hybrids have only been used for sensing and detecting the permissible and non-permissible levels of water pollutants rather than eliminating them from the complex waste water matrices. A few studies have shown effective degradation of various water pollutants using immobilized enzyme catalysts onto different supports [17, 34, 95]. So, there is a room for developing and optimizing a CNT-Enzyme (Nanobiohybrid) catalyst which would be a novel water purification technology suited for attenuating the waste water pollutants in the near future.

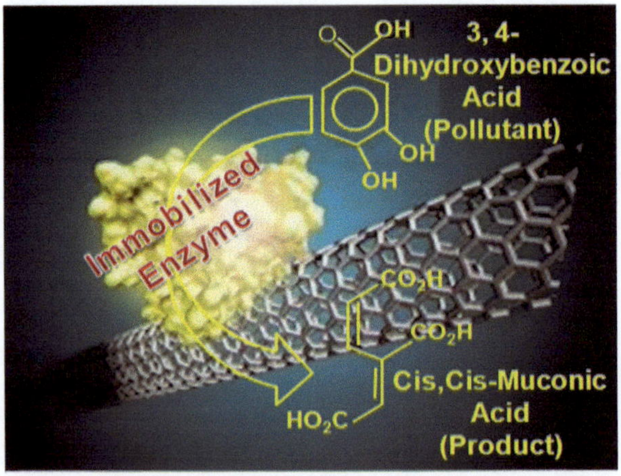

Fig. 1.6 Illustration of an immobilized enzyme onto CNT for pollutant degradation

1.5.1 *Protocatechuate 3,4-dioxygenase: an Enzyme for Water Treatment*

Protocatechuate 3,4-dioxygenase or simply 3,4-POD (EC 1.13.1.3) is an intradiol cleaving enzyme with a nonheme Fe (III) at the active site (Fig. 1.7) that catalyzes the oxidation of 3,4-DHBA [74] to 3-carboxy-cis,cis-muconate (Fig. 1.6) in presence of O_2 [60]. Thus, the two substrates of this enzyme are 3,4-DHBA, O_2, whereas its product is 3-carboxy-cis,cis-muconate.

The enzyme has been obtained from several sources, but extensively purified, crystallized, and characterized from *Pseudomonas putida* (*P. putida*) [27, 72]. A few studies have been adopted to immobilize 3,4-POD on porous glass [90, 91], agarose [34, 98] and alginate matrix [34, 71]. However, the studies have found without due scrutiny, since crude 3,4-POD extracts have been used [34, 71] that could mislead the observed catalytic efficiency. In addition, no characterizations and kinetic experiments have performed of immobilized 3,4-POD [34, 71, 91, 92, 98], leaving pitfall for potential scientist to explore more on this field.

1.5.2 *Protocatechuate 3,4-dioxygenase Immobilized on Carbon Nanotube*

The judicious choices that influence to select CNTs [18, 85] as support for immobilizing 3,4-POD in water treatments are (i) high surface area, (ii) large aspect ratio, (iii) porous architectures, (iv) high chemical reactivity, and (v) easy to functionalize and operate. These allow CNT as an excellent adsorbent material to trap various waste water pollutants. So, the material postulates synergistic effect in a way that it not only adsorb the water pollutants [84], but also acts as suitable support to hold enzymes for an effective biodegradation processes [25]. In addition, CNTs are stable and inert for microbial contaminations that might give immobilized 3,4-POD prolong activity, antifouling, self-cleaning, and reusable functions in waste water treatment plants. Finally, a hollow CNT could provide frictionless transport of water molecules, and this makes them suitable for the development of high fluxing separation techniques and would provide low energy solution for water treatment [17, 19]. However, general considerations for using MWCNTs are because of their low preparation cost and availability in large quantities in this project.

However, pristine CNTs (as received or synthesized) are often contaminated with various metal catalysts and non-nanotube carbonaceous agents such as amorphous carbons, ash, and so on [36]. Metal and amorphous carbon impurities could contribute toward the observed catalytic behavior and misleading total electrocatalytic activity [9]. Further, CNT impurities could have several detrimental effects in enzyme immobilization such as decreasing immobilization efficiency, enzyme's conformation changes, alteration of its catalytic activity, chelating, and so on. Metals in CNTs could inhibit total catalytic efficiency [49] of an immobilized

Fig. 1.7 Protocatechuate 3,4-dioxygenase hetero24mer + 3,4DHBA (*green*) + 12 Fe (l.*blue*) from *Pseudomonas putida*

enzyme. Therefore, purification of pristine CNTs is prerequisite before developing an effective enzyme immobilization system for waste water purifications. Another major hurdle is to manipulate CNT solubility in water solution. If a system used to eliminate pollutants out of water, one might bring water and/or pollutants to the reactive surfaces of the Nanobiohybrid catalyst. MWCNTs have shown insoluble in water because of its hydrophobic graphite sheet [69], which might result to minimize maximum enzyme loading onto CNT surfaces [75]. In addition, inert CNTs required chemical modifications to covalently attach molecules on its surface [28]. Therefore, before developing a nanobiohybrid to degrade 3,4-DHBA in water, pristine MWCNTs should be purified and functionalized not only to remove its generic impurities, but to solubilize them into water. Although some chemical, physical and a combination of both [36] purification methods have been adopted to remove CNT impurities and manipulate their solubility, they might degrade CNT, shortening and burning under extreme heat and excessive chemicals [20]. Such

CNTs might not be useful as suitable support matrix for 3,4-POD immobilization to yield an efficacious nanobiohybrid. Therefore, removing MWCNT impurities by keeping its integrity with high solubility percentage has remained one of the toughest tasks to fabricate new CNT-based water purification technologies.

1.6 What are the Problems that We Fixed Here?

Conventional waste water purification technologies such as Fenton methods [29, 66], adsorption [68], O_3/UV or H_2O_2/UV methods [11] and microbial degradation [13] have been faced difficulties to tackle 3,4-DHBA. Although the Nanobiohybrid has been popularly used to degrade wide range of water pollutants [45, 95], to date no study has yet been attempted or published to develop a MWCNT-3,4-POD (Nanobio) hybrid catalyst for degrading 3,4-DHBA in water. Besides, most of the existing methods to remove CNT impurities and hydrophobicity that have thwarted its success in water purifications [1, 9] are not suitable because they often degrade CNT native skeleton, yielding burned, shortened, curled, slowly dispersed and poorly soluble CNTs [36]. Such CNTs are not suitable for developing effective Nanobiohybrid catalysts [75], leaving pitfall for potential scientists to spotlight the issue. Developing a recrudescence and reusable Nanobiohybrid catalyst by dodging existing challenges could upgrade current water purification technologies that might ensure plethora of 3,4-DHBA free fresh water for all in the near future.

1.6.1 Purification of Pristine Carbon Nanotube

One of the major problems that has significantly hindered or limited CNT applications is its impurities with extreme hydrophobicity [36]. The existing CNT synthesis methods have added metal catalysts to increase CNT yield and reduce cost [47]. The level of these unwanted matters depends on the specific method used for CNT synthesis. Whatever might be the method of choice; impurities often hinder CNT performances and confuse the understanding of their original functionalities, limiting their applications in many important fields. For instance, impurities could determine nanotube's pore diameters, morphology, and have chance to influence or inhibit adsorption of water pollutants [1]; making them unsuitable to use in waste water treatment plants. Therefore, before developing a nanobiohybrid catalyst; purity and solubility of pristine MWCNTs should be checked carefully. This is essential because of avoiding ambiguities in experimental findings as well as to develop confidence on the designed nanobiohybrid catalyst to be used in the subsequent studies.

In order to purify CNTs, three classes of CNT purification methods have been adopted [36]. These are chemical (HCl, H_2O_2, O_3, H_2SO_4/HNO_3, HNO_3, $KMnO_4$, $(NH_4)_2S_2O_8$, $KMnO_4/H_2SO_4$, etc.), physical (filtration, centrifugations, and high temperature annealing) and a combination of both [36]. Chemical purification methods have been observed effective because of its selectivity, sensitivity, faster rate kinetics of the oxidation of carbonaceous impurities and metal catalysts dissolution from the CNTs [36]. In addition, chemical agents are widely available, cost effective and need simple laboratory settings [86]. In contrast, physical methods have been involved to remove graphitic sheets and carbon nanospheres from CNTs [36]. The method has found ineffective to remove carbon impurities, complicated and time consuming [36]. Therefore, we developed simple chemical purification methods for yielding highly pure and intact MWCNTs with other desired properties that were necessary for designing a suitable platform for 3,4-POD immobilization.

1.6.2 Solubilization of Carbon Nanotube

Pristine CNTs are hydrophobic and have agglomerated and precipitated in water solvents rapidly [69]. But CNTs with surface functionalities could be soluble in water and brought CNTs to repulse each other in a solution [99]. This allows individual CNT that can travel freely into the solution avoiding aggregation in a homogenous solution. It would accelerate the affinity of interactions among water pollutants and biomolecules with CNT surfaces. To enable CNTs purification at individual scale and manipulate their solubility, various covalent and non-covalent functionalization strategies have been adopted [101]. In covalent modification, hydrophilic substituent has been introduced using various wet chemical treatments such as acidic, basic and oxidizing agents. Both tip end and sidewall functionalizations [15, 76] have been performed to reduce or eliminate hydrophobicity, and increase solubility and colloidal dispersibility [26]. In non-covalent modification, surfactant wrapping has widely been used to increase CNT solubility in different aqueous media [57]. Both types of the above approaches could change nanotubes' surface charges, and energy potentials, which alter their intra and/or inter molecular interactions and surface properties. However, covalent modifications have been observed better efficiency to introduce specific functional group necessary for enzyme immobilization [28]. For example, introducing carboxylic groups (–COOH) onto CNT surfaces has improved their water solubility and dispersion [69, 73]. The group has played critical point of enzyme and/or protein attachments on CNTs [28] via diimide-activated amidation reaction [37]. Consequently, we envisioned here to decorate MWCNTs with -COOH functionality that not only helped to improve MWCNT solubility, but to immobilize 3,4-POD onto MWCNT surfaces.

1.6.3 Development and Characterizations of Nanobiohybrid Catalyst

Enzymes could be immobilized on CNTs by covalent and non-covalent approaches [25]. Covalent bonding of enzymes with CNT has been popular by inducing the reaction of the free amine groups on the surface of an enzyme with –COOH group's activation using carbodiimide chemistry [5, 37, 43]. Linking molecules such as 1-pyrenebutanoic acid succinimidyl ester and poly (ethylene glycol) (PEG) have been used for covalent immobilization of enzymes such as HRP and perhydrolase S54V, respectively [21, 46]. On the other hand, non-covalent enzyme immobilization have dealt with physical adsorption, surfactant wrapping, layer-by-layer (LBL) attachments and so on [25]. Non-covalent adsorption of enzymes into CNTs has shown to occur by inheritance hydrophobic, electrostatic and π–π stacking interactions between CNTs and enzymes [28, 56].

However, appropriate immobilization techniques should be selected based on desired application fields where a Nanobiohybrid will be implemented. For instance, most of the non-covalent approaches have been popular for developing a hybrid catalyst in sensor technologies [25, 28]. This is because of retaining both enzymes and CNT native structures. But the major shortcoming of this approach is that the immobilized enzymes could loss gradually from the system for long time uses [28]. In contrast, covalent attachment could be more preferable for a strong and durable CNT-enzyme attachment at operations [25]. As a result, here we foresaw the advantages of covalent immobilization of 3,4-POD onto MWCNTs. No matter what immobilization techniques have been chosen as long as one could maintain stable enzyme's attachment onto supports and improve its stability and functions to their free state [61].

Characterizing a developed technology plays vital roles for improving its productivity which indicates that the designed technology is maturing. In reality, a technology cannot be fulfilled without seeing its physical properties and actions in applied field. For instance, the theory of relativity in physics by Albert Einstein [22] has been popular because of its accurate validation and great significance in various fields. However, MWCNTs will be characterized by transmission electron microscope (TEM), Energy-dispersive X-ray spectroscope (EDX), attenuated total reflectance infrared (ATR-IR) spectroscope, X-ray photoelectron spectroscope (XPS), Raman spectroscope, Thermogravimetric analysis (TGA) and Ultraviolet–visible (UV/vis) spectrophotometric studies [36, 51]. In addition, Nanobiohybrid catalyst will be characterized using scanning electron microscope (SEM), TEM, atomic force microscope (AFM), ATR-IR, UV/vis spectroscope, and circular dichroism (CD) studies [75, 95].

1.6.4 Nanobiohybrid Catalyst for Pollutant Removal

Nanobiohybrid catalyst's performances will be evaluated by degrading of 3,4-DHBA in artificial waste water through optimized conditions. Both the catalytic efficiency and stabilities of the Nanobiohybrid will be experimentally and statistically validated for comparing both the free and immobilized 3,4-POD activities. Once it gets mature and shows high stability and good catalytic behaviors, then it could be used in future to eliminate 3,4-DHBA or its derivative water pollutants from different waste water matrices, and hopefully would be a sustainable approach to manage plethora fresh water for all.

1.7 Conclusions

The development of a useful, stable, and efficacious MWCNT-3.4-POD (Nanobio)-hybrid catalyst for degrading a notorious phenol-type water pollutant called 3,4-DHBA has been reported. The chapter reveals that the utilization of useful CNT's properties (e.g., high surface area, greater chemical reactivity, lower costs and energy, and antimicrobial activity) for the development of an enzyme-based Nanobiohybrid catalytic system for water purification. Herein efforts are given to decorate CNT as suitable support matrix for 3,4-POD immobilization. The Nanobiohybrid not only proved to decontaminate 3,4-DHBA, but it would help understanding of an enzyme behavior at nanointerfaces. Therefore, it has appeals to register as a novel approach not only for water purification, but to expand Ab ovo applications.

References and Future Readings

1. Agnihotri, S., Mota, J.P., Rostam-Abadi, M., Rood, M.J.: Adsorption site analysis of impurity embedded single-walled carbon nanotube bundles. Carbon **44**(12), 2376–2383 (2006)
2. Ali, M., Das, R., Maamor, A., Hamid, S.B.A.: Multifunctional Carbon Nanotubes (CNTs): a New Dimension in Environmental Remediation. Adv. Mater. Res. **832**, 328–332 (2014)
3. Aragay, G., Pons, J., Merkoçi, A.: Recent trends in macro-, micro-, and nanomaterial-based tools and strategies for heavy-metal detection. Chem. Rev. **111**(5), 3433–3458 (2011)
4. Asuri, P., Bale, S.S., Karajanagi, S.S., Kane, R.S.: The protein–nanomaterial interface. Curr. Opin. Biotechnol. **17**(6), 562–568 (2006)
5. Asuri, P., Karajanagi, S.S., Sellitto, E., Kim, D.Y., Kane, R.S., Dordick, J.S.: Water-soluble carbon nanotube-enzyme conjugates as functional biocatalytic formulations. Biotechnol. Bioeng. **95**(5), 804–811 (2006)
6. Babich, H., Sedletcaia, A., Kenigsberg, B.: In vitro cytotoxicity of protocatechuic acid to cultured human cells from oral tissue: involvement in oxidative stress. Pharmacol. Toxicol. **91**(5), 245–253 (2002)
7. Baker, R., Waite, R.: Formation of carbonaceous deposits from the platinum-iron catalyzed decomposition of acetylene. J. Catal. **37**(1), 101–105 (1975)
8. Baker, R., Barber, M., Harris, P., Feates, F., Waite, R.: Nucleation and growth of carbon deposits from the nickel catalyzed decomposition of acetylene. J. Catal. **26**(1), 51–62 (1972)
9. Banks, C.E., Crossley, A., Salter, C., Wilkins, S.J., Compton, R.G.: carbon nanotubes contain metal impurities which are responsible for the "electrocatalysis" seen at some nanotube-modified electrodes. Angew. Chem. Int. Ed. **45**(16), 2533–2537 (2006)
10. Belin, T., Epron, F.: Characterization methods of carbon nanotubes: a review. Mater. Sci. Eng. B **119**(2), 105–118 (2005)
11. Benitez, F.J., Beltran-Heredia, J., Acero, J.L., Gonzalez, T.: Degradation of protocatechuic acid by two advanced oxidation processes: ozone/UV radiation and H_2O_2UV radiation. Water Res. **30**(7), 1597–1604 (1996)
12. Brena, B., González-Pombo, P., Batista-Viera, F.: Immobilization of Enzymes: A Literature Survey. In: Immobilization of Enzymes and Cells, pp. 15–31. Springer, Berlin (2013)
13. Buchan, A., Collier, L.S., Neidle, E.L., Moran, M.A.: Key aromatic-ring-cleaving enzyme, protocatechuate 3,4-dioxygenase, in the ecologically important marineRoseobacter lineage. Appl. Environ. Microbiol. **66**(11), 4662–4672 (2000)
14. Canela, M., Jardim, W.: Identification and photocatalytic destruction of malodorous compounds in sewage. Environ. Technol. **29**(6), 673–679 (2008)
15. Chen, J., Hamon, M.A., Hu, H., Chen, Y., Rao, A.M., Eklund, P.C., Haddon, R.C.: Solution properties of single-walled carbon nanotubes. Science **282**(5386), 95–98 (1998)

16. Coleridge, S.T.: The rime of the ancient mariner and other poems, vol. 80. Houghton, Mifflin (1895)
17. Dai, Y., Yin, L., Niu, J.: Laccase-carrying electrospun fibrous membranes for adsorption and degradation of PAHs in shoal soils. Environ. Sci. Technol. **45**(24), 10611–10618 (2011)
18. Das, R., Abd Hamid, S.B., Ali, M.E., Ismail, A.F., Annuar, M.S.M., Ramakrishna, S.: Multifunctional carbon nanotubes in water treatment: the present, past and future. Desalination **354**, 160–179 (2014). doi:10.1016/j.desal.2014.09.032
19. Das, R., Ali, M.E., Abd Hamid, S.B., Ramakrishna, S., Chowdhury, Z.Z.: Carbon nanotube membranes for water purification: a bright future in water desalination. Desalination **336**, 97–109 (2014). doi:10.1016/j.desal.2013.12.026
20. Das, R., Ali, M.E., Bee Abd Hamid, S., Annuar, M.S.M., Ramakrishna, S.: Common wet chemical agents for purifying multiwalled carbon nanotubes. J. Nanomaterials **2014**, 9 (2014). doi:10.1155/2014/945172
21. Dinu, C.Z., Zhu, G., Bale, S.S., Anand, G., Reeder, P.J., Sanford, K., Whited, G., Kane, R. S., Dordick, J.S.: Enzyme-based nanoscale composites for use as active decontamination surfaces. Adv. Funct. Mater. **20**(3), 392–398 (2010)
22. Einstein A (1920) Relativity: the special and general theory. Penguin, Harmondsworth
23. Fathizadeh, M., Aroujalian, A., Raisi, A.: Effect of added NaX nano-zeolite into polyamide as a top thin layer of membrane on water flux and salt rejection in a reverse osmosis process. J. Membr. Sci. **375**(1), 88–95 (2011)
24. Filleter, T., Bernal, R., Li, S., Espinosa, H.D.: Ultrahigh strength and stiffness in cross-linked hierarchical carbon nanotube bundles. Adv. Mater. **23**(25), 2855–2860 (2011)
25. Feng, W., Ji, P.: Enzymes immobilized on carbon nanotubes. Biotechnol. Adv. **29**(6), 889–895 (2011)
26. Fogden, S., Verdejo, R., Cottam, B., Shaffer, M.: Purification of single walled carbon nanotubes: the problem with oxidation debris. Chem. Phys. Lett. **460**(1), 162–167 (2008)
27. Fujisawa, H., Hayaishi, O.: Protocatechuate 3,4-dioxygenase I. Crystallization and characterization. J. Biol. Chem. **243**(10), 2673–2681 (1968)
28. Gao, Y., Kyratzis, I.: Covalent immobilization of proteins on carbon nanotubes using the cross-linker 1-ethyl-3-(3-dimethylaminopropyl) carbodiimide—a critical assessment. Bioconjug. Chem **19**(10), 1945–1950 (2008)
29. Gernjak, W., Krutzler, T., Glaser, A., Malato, S., Caceres, J., Bauer, R., Fernández-Alba, A.: Photo-Fenton treatment of water containing natural phenolic pollutants. Chemosphere **50**(1), 71–78 (2003)
30. Girelli, A.M., Mattei, E., Messina, A.: Phenols removal by immobilized tyrosinase reactor in on-line high performance liquid chromatography. Anal. Chim. Acta **580**(2), 271–277 (2006)
31. Goh, P., Ismail, A., Ng, B.: Carbon nanotubes for desalination: performance evaluation and current hurdles. Desalination **308**, 2–14 (2013)
32. Guiseppi-Elie, A., Lei, C., Baughman, R.H.: Direct electron transfer of glucose oxidase on carbon nanotubes. Nanotechnology **13**(5), 559 (2002)
33. Gupta, V.K., Ali, I., Saleh, T.A., Nayak, A., Agarwal, S.: Chemical treatment technologies for waste-water recycling—an overview. RSC Adv. **2**(16), 6380–6388 (2012)
34. Guzik, U., Hupert-Kocurek, K., Krysiak, M., Wojcieszyńska, D.: Degradation potential of protocatechuate 3,4-dioxygenase from crude extract of *Stenotrophomonas maltophilia* strain KB2 immobilized in calcium alginate hydrogels and on glyoxyl agarose. BioMed Research International 2014 (2014)
35. Hamdi, M.: Toxicity and biodegradability of olive mill wastewaters in batch anaerobic digestion. Appl. Biochem. Biotechnol. **37**(2), 155–163 (1992)
36. Hou, P.-X., Liu, C., Cheng, H.-M.: Purification of carbon nanotubes. Carbon **46**(15), 2003–2025 (2008)
37. Huang, W., Taylor, S., Fu, K., Lin, Y., Zhang, D., Hanks, T.W., Rao, A.M., Sun, Y.-P.: Attaching proteins to carbon nanotubes via diimide-activated amidation. Nano Lett. **2**(4), 311–314 (2002)
38. Hong, S., Myung, S.: Nanotube electronics: a flexible approach to mobility. Nat. Nanotechnol. **2**(4), 207–208 (2007)

39. Iijima, S.: Helical microtubules of graphitic carbon. nature **354**(6348), 56–58 (1991)
40. Iijima, S., Ichihashi, T.: Single-shell carbon nanotubes of 1-nm diameter. Nature **363**(6430), 603–605 (1993). doi:10.1038/363603a0
41. Jadav, G.L., Singh, P.S.: Synthesis of novel silica-polyamide nanocomposite membrane with enhanced properties. J. Membr. Sci. **328**(1), 257–267 (2009)
42. Jeong, B.-H., Hoek, E., Yan, Y., Subramani, A., Huang, X., Hurwitz, G., Ghosh, A.K., Jawor, A.: Interfacial polymerization of thin film nanocomposites: a new concept for reverse osmosis membranes. J. Membr. Sci. **294**(1), 1–7 (2007)
43. Jiang, K., Schadler, L.S., Siegel, R.W., Zhang, X., Zhang, H., Terrones, M.: Protein immobilization on carbon nanotubes via a two-step process of diimide-activated amidation. J. Mater. Chem. **14**(1), 37–39 (2004)
44. Kamat, P.V., Meisel, D.: Nanoscience opportunities in environmental remediation. C. R. Chim. **6**(8), 999–1007 (2003)
45. Khin, M.M., Nair, A.S., Babu, V.J., Murugan, R., Ramakrishna, S.: A review on nanomaterials for environmental remediation. Energy Environ. Sci. **5**(8), 8075–8109 (2012)
46. Kim, B.J., Kang, B.K., Bahk, Y.Y., Yoo, K.H., Lim, K.J.: Immobilization of horseradish peroxidase on multi-walled carbon nanotubes and its enzymatic stability. Curr. Appl. Phys. **9** (4), e263–e265 (2009)
47. Kobayashi, Y., Nakashima, H., Takagi, D., Homma, Y.: CVD growth of single-walled carbon nanotubes using size-controlled nanoparticle catalyst. Thin Solid Films **464**, 286–289 (2004)
48. Kolaczkowski, S., Beltran, F., McLurgh, D., Rivas, F.: Wet air oxidation of phenol: factors that may influence global kinetics. Process Saf. Environ. Prot. **75**(4), 257–265 (1997)
49. Latha, R., Mandappa, I., Thakur, M., Manonmani, H.: Influence of metal ions on dehydrohalogenase activity. Afr. J. Basic Appl. Sci. **3**(2), 45–51 (2011)
50. Lee, Y.-M., Kwon, O.-Y., Yoon, Y.-J., Ryu, K.: Immobilization of horseradish peroxidase on multi-wall carbon nanotubes and its electrochemical properties. Biotechnol. Lett. **28**(1), 39–43 (2006)
51. Lehman, J.H., Terrones, M., Mansfield, E., Hurst, K.E., Meunier, V.: Evaluating the characteristics of multiwall carbon nanotubes. Carbon **49**(8), 2581–2602 (2011)
52. Liu, Y., Qu, X., Guo, H., Chen, H., Liu, B., Dong, S.: Facile preparation of amperometric laccase biosensor with multifunction based on the matrix of carbon nanotubes–chitosan composite. Biosens. Bioelectron. **21**(12), 2195–2201 (2006)
53. Nakamura, Y., Torikai, K., Ohigashi, H.: Toxic dose of a simple phenolic antioxidant, protocatechuic acid, attenuates the glutathione level in ICR mouse liver and kidney. J. Agric. Food Chem. **49**(11), 5674–5678 (2001)
54. Nakamura, Y., Torikai, K., Ohto, Y., Murakami, A., Tanaka, T., Ohigashi, H.: A simple phenolic antioxidant protocatechuic acid enhances tumor promotion and oxidative stress in female ICR mouse skin: dose-and timing-dependent enhancement and involvement of bioactivation by tyrosinase. Carcinogenesis **21**(10), 1899–1907 (2000)
55. Nelson, D.L., Lehninger, A.L., Cox, M.M.: Lehninger principles of biochemistry. Macmillan, NY (2008)
56. Nepal, D., Geckeler, K.E.: pH-sensitive dispersion and debundling of single-walled carbon nanotubes: lysozyme as a tool. Small **2**(3), 406–412 (2006)
57. O'connell, M.J., Bachilo, S.M., Huffman, C.B., Moore, V.C., Strano, M.S., Haroz, E.H., Rialon, K.L., Boul, P.J., Noon, W.H., Kittrell, C.: Band gap fluorescence from individual single-walled carbon nanotubes. Science **297**(5581), 593–596 (2002)
58. Oberrecht, K.: The effects of rising sea levels. 2014 (4) (2014)
59. Ohta, K., Nishizawa, T., Nishiguchi, T., Shimizu, R., Hattori, Y., Inoue, S., Katayama, M., Mizu-Uchi, K., Kono, T.: Synthesis of carbon nanotubes by microwave heating: Influence of diameter of catalytic Ni nanoparticles on diameter of CNTs. J. Mater. Chem. A **2**(8), 2773–2780 (2014)
60. Ornston, L., Stanier, R.: The conversion of catechol and protocatechuate to β-ketoadipate by *Pseudomonas putida* I. Biochemistry. J. Biol. Chem. **241**(16), 3776–3786 (1966)

61. Pedrosa, V.A., Paliwal, S., Balasubramanian, S., Nepal, D., Davis, V., Wild, J., Ramanculov, E., Simonian, A.: Enhanced stability of enzyme organophosphate hydrolase interfaced on the carbon nanotubes. Colloids Surf. B **77**(1), 69–74 (2010)
62. Poulios, I., Makri, D., Prohaska, X.: Photocatalytic treatment of olive milling waste water: oxidation of protocatechuic acid. Glob. Nest: Int J **1**, 55–62 (1999)
63. Prasek, J., Drbohlavova, J., Chomoucka, J., Hubalek, J., Jasek, O., Adam, V., Kizek, R.: Methods for carbon nanotubes synthesis-review. J. Mater. Chem. **21**(40), 15872–15884 (2011). doi:10.1039/c1jm12254a
64. Ramani, K., Karthikeyan, S., Boopathy, R., Kennedy, L.J., Mandal, A., Sekaran, G.: Surface functionalized mesoporous activated carbon for the immobilization of acidic lipase and their application to hydrolysis of waste cooked oil: isotherm and kinetic studies. Process Biochem. **47**(3), 435–445 (2012)
65. Reznik, D., Olk, C., Neumann, D., Copley, J.: X-ray powder diffraction from carbon nanotubes and nanoparticles. Phys. Rev. B **52**(1), 116 (1995)
66. Rivas, F.J., Frades, J., Alonso, M.A., Montoya, C., Monteagudo, J.M.: Fenton's oxidation of food processing wastewater components. Kinetic modeling of protocatechuic acid degradation. J. Agric. Food Chem. **53**(26), 10097–10104 (2005). doi:10.1021/jf0512712
67. San, N., Hatipoğlu, A., Koçtürk, G., Çınar, Z.: Prediction of primary intermediates and the photodegradation kinetics of 3-aminophenol in aqueous TiO_2 suspensions. J. Photochem. Photobiol. A **139**(2), 225–232 (2001)
68. Sarma, J., Mahiuddin, S.: Specific ion effect on the point of zero charge of α-alumina and on the adsorption of 3,4-dihydroxybenzoic acid onto α-alumina surface. Colloids Surf. A **457**, 419–424 (2014). doi:10.1016/j.colsurfa.2014.06.014
69. Shieh, Y.-T., Liu, G.-L., Wu, H.-H., Lee, C.-C.: Effects of polarity and pH on the solubility of acid-treated carbon nanotubes in different media. Carbon **45**(9), 1880–1890 (2007)
70. Siddique, M.H., St Pierre, C.C., Biswas, N., Bewtra, J.K., Taylor, K.E.: Immobilized enzyme catalyzed removal of 4-chlorophenol from aqueous solution. Water Res. **27**(5), 883–890 (1993)
71. Silva, ASd, Jacques, R.J.S., Andreazza, R., Bento, F.M., Camargo, FAdO: The effects of trace elements, cations, and environmental conditions on protocatechuate 3,4-dioxygenase activity. Scientia Agricola **70**(2), 68–73 (2013)
72. Sim, H.W., Jung, M., Cho, Y.K.: Purification and characterization of protocatechuate 3, 4-dioxygenase from Pseudomonas pseudoalcaligenes KF707. J. Korean Soc. Appl. Biol. Chem. **56**(4), 401–408 (2013)
73. Smith, B., Wepasnick, K., Schrote, K.E., Cho, H.-H., Ball, W.P., Fairbrother, D.H.: Influence of surface oxides on the colloidal stability of multi-walled carbon nanotubes: A structure—property relationship. Langmuir **25**(17), 9767–9776 (2009)
74. Stanier, R., Ingraham, J.: Protocatechuic acid oxidase. J. Biol. Chem. **210**(2), 799–808 (1954)
75. Subrizi, F., Crucianelli, M., Grossi, V., Passacantando, M., Pesci, L., Saladino, R.: Carbon nanotubes as activating tyrosinase supports for the selective synthesis of catechols. ACS Catal. **4**(3), 810–822 (2014)
76. Sun, Y.-P., Fu, K., Lin, Y., Huang, W.: Functionalized carbon nanotubes: properties and applications. Acc. Chem. Res. **35**(12), 1096–1104 (2002)
77. Takagi, Y., Tauchi, L., Nguyen-Tran, H.-D., Ohta, T., Shimizu, M., Ohta, K.: Development of a novel method to synthesize carbon nanotubes from granulated polystyrene and nickel nanoparticles by microwave heating. J. Mater. Chem. **21**(38), 14569–14574 (2011)
78. Tan, H., Feng, W., Ji, P.: Lipase immobilized on magnetic multi-walled carbon nanotubes. Bioresour. Technol. **115**, 172–176 (2012)
79. Tanaka, T., Kojima, T., Kawamori, T., Yoshimi, N., Mori, H.: Chemoprevention of diethylnitrosamine-induced hepatocarcinogenesis by a simple phenolic acid protocatechuic acid in rats. Can. Res. **53**(12), 2775–2779 (1993)
80. UN-News (2008) Ban Ki-moon warns that water shortages are increasingly driving conflicts

81. UN (2013) The Report of the High-Level Panel of Eminent Persons on the Post-2015 Development Agenda. UN, New York
82. UN (2014) Water scarcity. UN. http://www.un.org/waterforlifedecade/scarcity.shtml. Accessed 4, September 2014
83. UNESCAP: Building resilience to natural disasters and major economic crises. UNESCAP, Bangkok (2013)
84. Upadhyayula, V.K.K., Deng, S.G., Mitchell, M.C., Smith, G.B.: Application of carbon nanotube technology for removal of contaminants in drinking water: a review. Sci. Total Environ. **408**(1), 1–13 (2009). doi:10.1016/j.scitotenv.2009.09.027
85. Verma, M.L., Naebe, M., Barrow, C.J., Puri, M.: Enzyme Immobilisation on amino-functionalised multi-walled carbon nanotubes: structural and biocatalytic characterisation. PLoS ONE **8**(9), e73642 (2013)
86. Wepasnick, K.A., Smith, B.A., Schrote, K.E., Wilson, H.K., Diegelmann, S.R., Fairbrother, D.H.: Surface and structural characterization of multi-walled carbon nanotubes following different oxidative treatments. Carbon **49**(1), 24–36 (2011)
87. WHO (2002) World health report: Reducing risks, promoting healthy life. WHO. http://www.who.int/whr/2002/en/whr02_en.pdf. Accessed 5 Sept 2014
88. WHO, UNICEF.: Global water supply and sanitation assessment, 2000 report. http://www.who.int/water_sanitation_health/monitoring/jmp2000.pdf (2000)
89. WHO, UNICEF.: Progress on sanitation and drinking-water-2013 update: joint monitoring programme for water supply and sanitation (2013)
90. Wojtaś-Wasilewska, M., Luterek, J., Rogalski, J.: Immobilization of protocatechuate 3, 4-dioxygenase from *Pleurotus ostreatus* on activated porous glass beads. Phytochemistry **27** (9), 2731–2733 (1988)
91. Wojtaś-Wasilewska, M., Luterek, J., Leonowicz, A., Dawidowicz, A.: Dearomatization of lignin derivatives by fungal protocatechuate 3,4-dioxygenase immobilized on porosity glass. Biotechnol. Bioeng. **32**(4), 507–511 (1988)
92. WWAP: The United Nations World Water Development Report 3: Water in a Changing World. UNESCO/Earthscan, Paris/London (2009)
93. WWAP: The United Nations World Water Development Report 4: Managing Water Under Uncertainty and Risk. UNESCO, Paris (2012)
94. Wang, Z., Ci, L., Chen, L., Nayak, S., Ajayan, P.M., Koratkar, N.: Polarity-dependent electrochemically controlled transport of water through carbon nanotube membranes. Nano Lett. **7**, 697–702 (2007)
95. Xu, R., Chi, C., Li, F., Zhang, B.: Laccase-polyacrylonitrile nanofibrous membrane: highly immobilized, stable, Reusable, and Efficacious for 2,4,6-trichlorophenol removal. ACS Appl. Mater. Interfaces. **5**(23), 12554–12560 (2013)
96. Yang, H.Y., Han, Z.J., Yu, S.F., Pey, K.L., Ostrikov, K., Karnik, R. Carbon nanotube membranes with ultrahigh specific adsorption capacity for water desalination and purification. Nature Commun. **4**, 2220 (2013)
97. Yoshimura, S., Chang, R.P.H.: Supercarbon: synthesis, properties and applications, vol. 33. Springer, Berlin (1998)
98. Zaborsky, O.R., Ogletree, J.: Immobilization of protocatechuate 3,4-dioxygenase with activated agarose. Biochimica et Biophysica Acta (BBA)-Enzymology 289 **1**, 68–76 (1972)
99. Zeng, L., Zhang, L., Barron, A.R.: Tailoring aqueous solubility of functionalized single-wall carbon nanotubes over a wide pH range through substituent chain length. Nano Lett. **5**(10), 2001–2004 (2005)
100. Zhang, G., Ma, J., Wang, J., Li, Y., Zhang, G., Zhang, F., Fan, X.: Lipase immobilized on graphene oxide as reusable biocatalyst. Ind. Eng. Chem. Res. **53**(51), 19878–19883 (2014)
101. Zhang, M., Su, L., Mao, L.: Surfactant functionalization of carbon nanotubes (CNTs) for layer-by-layer assembling of CNT multi-layer films and fabrication of gold nanoparticle/CNT nanohybrid. Carbon **44**(2), 276–283 (2006)
102. Zhao, Y.L., Stoddart, J.F.: Noncovalent functionalization of single-walled carbon nanotubes. Acc. Chem. Res. **42**(8), 1161–1171 (2009). doi:10.1021/ar900056z

Chapter 2
Carbon Nanotube in Water Treatment

Thousands have lived without love, not one without water.
—Source: W.H. Auden: *Collected Poems: Auden*
by W.H. Auden, 1991.

Abstract The availability of safe and clean water is decreasing day by day, which is expected to increase in upcoming decades. To address this problem, various water purification technologies have been adopted. Among the various concepts proposed, CNTs based water treatment technologies have found to be promising because of its large surface area, high aspect ratio, greater chemical reactivity, lower cost, and energy, less chemical mass and impact on the environment. Therefore, research development and commercial interests in CNT are growing worldwide to treat water contaminants, which have huge impacts on the entire living systems including terrestrial, aquatic, and aerial flora and fauna. Here we reviewed most of the effective CNT based water purification technologies such as adsorption, hybrid catalysis, desalination, disinfection, sensing and monitoring of three major classes such as organic, inorganic and biological water pollutants. Since the Nanobiohybrid field yet remains to be matured, special importance has been paid on its mediated water purification technology. We have forayed into the deeper thoughts and compiled promises, facts and challenges of the important water purification technologies. Since water purification is a complex process; hydrologists, membrane technologists, environmentalists and industrialists can design "ONE POT" combination where effective water purification technologies would instate to tackle both the conventional and newly emerging toxic pollutants effectively.

2.1 Background

Earth water—a generous gift from the almighty God, and is the major constituent for all living organisms. Water has covered around 71% of the Earth's surface area [1]. Figure 2.1 shows three pie charts of total water volume, and its distributions on

R. Das, *Nanohybrid Catalyst based on Carbon Nanotube*,
Carbon Nanostructures, DOI 10.1007/978-3-319-58151-4_2

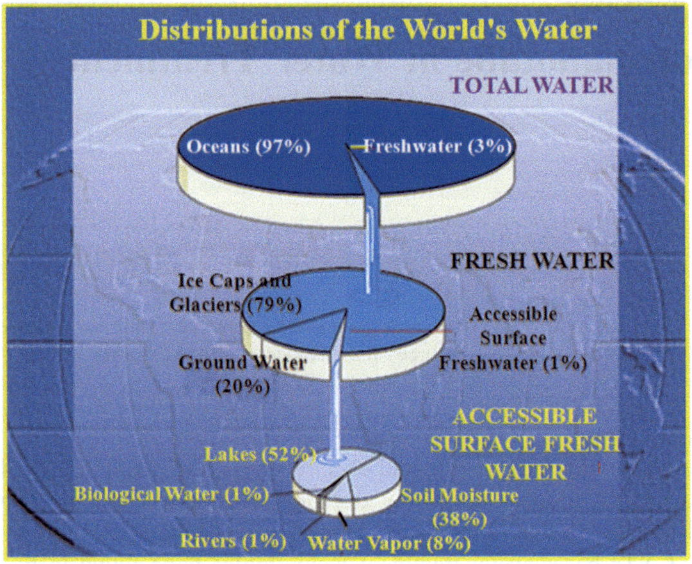

Fig. 2.1 Pie charts of total water volume on, in and above the Earth. The figure is adapted with permission from the Earth-Forum [25]

Earth's surfaces [25]. According to Fig. 2.1, water can exist in different forms such as gaseous (cloud vapors and atmospheres), liquids (seas and oceans, rivers, lakes), and solid waters (ice and glaciers). Although a higher percentage of water exists as saline water in the sea and oceans (97%) (Fig. 2.1 top chart), it cannot be used for drinking purpose due to its salinity and impurities.

Only (3%) freshwater exists as 79, 20 and 1% of the frozen, ground and accessible fresh surface waters, respectively (Fig. 2.1 middle chart). This accessible fresh water (1%) covers 53% in the rivers and streams; and the remaining percentages are in the forms of vapor, moisture, and biological (Fig. 2.1 down chart). These clearly indicate the unavailability of pure fresh water facilities for drinking which are necessary for sustaining and growing of all living organisms on Earth.

While limited fresh water availability is so obvious, many persistent pollutants have been simultaneously added into the world water bodies through both point and nonpoint origins. This leads to a global decrease in the availability of clean and safe water forms of water resources. Three major classes of water pollutants such as organic, inorganic, and biological contaminants have been dwindled the finite fresh water resources. Moreover, climate change due to persistent global warming could bring variation in natural systems, leading to ice-melting, sea level rise, soil, and fresh water submergence, increase evaporation and so on, which have collectively been contaminated the oceans [21]. While it is difficult to control or significantly reduce water pollution, the lack of cost-effective water purification technology has accelerated the problems. At some places, polluted water is not treatable for recycling especially in developing countries because of lacking cost effective, but sensitive water purification technology. Therefore, a novel, sensitive and cost-effective water purification technology is an urgent need, and the failure to develop such a system might further endanger the life processes and eco-friendly human existence.

2.2 Water Scarcity and Its Consequences

Increasing world population with water demands is highly interlinked and interdependent with each other. World population has been increasing day by day (80 million/year) [135, 136], which could directly increase water demand in the near future. Figure 2.2 represents the trends of population growths and their water demands from 1950 to 2025. This alarms the looming crisis of water can be seen, if corrective steps are not taken at right time.

Figure 2.3 represents the water scarcity of different areas on Earth. It shows many countries have encountered with little, physical and economical water scarcities, while some are waiting to reach. In the last century, water usages have been more than twice the rate of population growth [129], resulting in global water shortages. Such fresh water crisis brings problems not only to human, but also to the eco-systems. It has already been affected many continents, mainly developing countries, which accommodate almost 40% populace of our planet [129]. People in the developing countries depend mainly on the ground water facility due to the lack of expensive water purification techniques. Approximately 30% of the fresh and clean water accumulated in aquifers have commonly been utilized for irrigation to cultivate food grains, vegetables and fruits, leading to a water depletion of about 12 billion m^3/year [58], endangering biodiversities, flora and fauna.

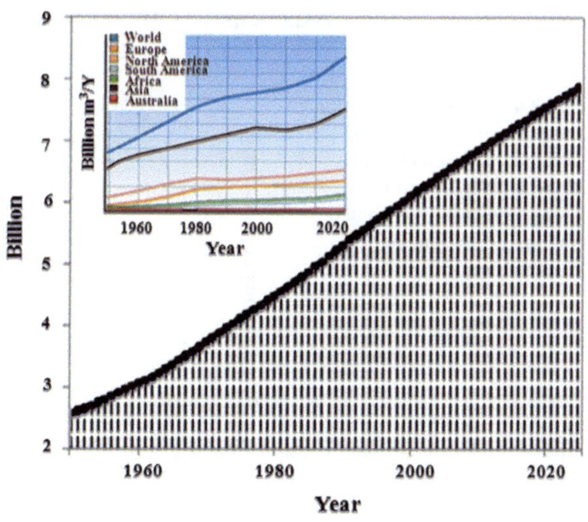

Fig. 2.2 Schematic representation of population growth (*outer plot*) and their corresponding water demand (*inner plot*) from 1950 to 2025. The plot was generated using the data from [58, 132]

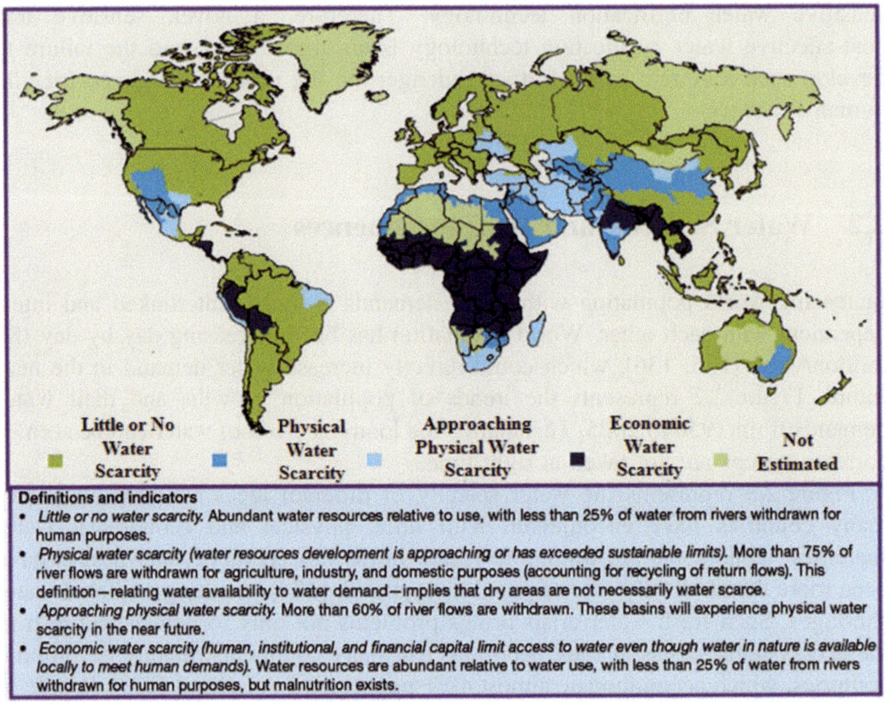

Fig. 2.3 Areas encountered with water scarcities on Earth. The figure is adapted with permission from the International Water Management Institute (IWMI) [49]

Water crisis has directly linked to the global food, and health security, economic growth, social progress and ultimately political stability [59]. Figure 2.4 demonstrates a relationship between gross domestic product (GDP) and the rate of water consumptions in some Asian cities [130]. Although per capita GDP of Singapore,

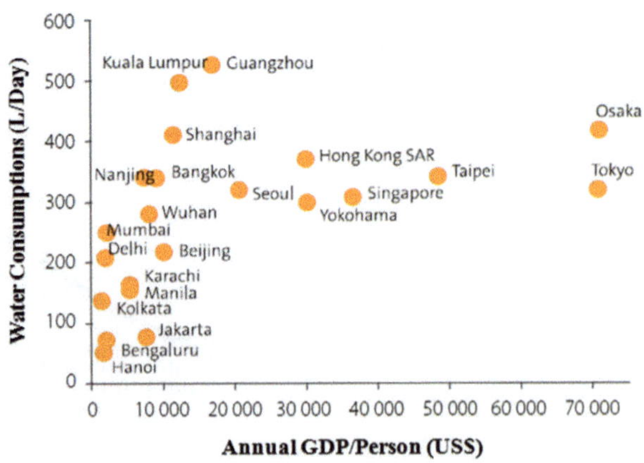

Fig. 2.4 Water consumption and per capita income in selected Asian cities. The figure is adapted with permission from the United Nations Development Programme [130]

Taipei and Tokyo have found more than Kuala Lumpur, Guangzhou, Shanghai and Hong Kong SAR; its per capita water consumptions are considerably lower. It hypothesizes a proportional relationship between water demands and per capita growth. The cities that are rapidly developing total water demands, which simultaneously create an inhibitory effect to the growth of per capita also.

Accumulating wastewaters from surroundings has been increased the transmission of various illnesses especially mosquito, water, and food-borne diseases such as dengue, malaria and hepatitis. Different water pollutants have shown different detrimental human health effects. For instances, persistent organic pollutants (POPs) [145] cause hormonal imbalance, alter reproductive behavior and birth defects, cancer, heart diseases and so on [110]. These have shown harmful effects because of their desensitized behaviors to chemical, biological, and photocatalytic degradation processes [110]. Consequently, they could have long-range transport, high cellular uptake followed by bioaccumulation into the human body that leads to cellular toxicity [110].

First, the most notorious organic wastewater pollutants are phenols and its derivatives which have been persisted more commonly in nature and are toxic for biological bodies [34]. Most of them have been leached and mixed into the water environment from both natural and anthropogenic sources. For instance, they often come up with the pyrolysis of forest trees and volcanic eruptions. Aquatic flora and fauna such as sweet flag and algae have known to synthesize many phenolic compounds as secondary metabolites [37, 83]. In addition, plant's and animal's decompositions have acted as natural sources for phenol-type water pollutants [34]. However, anthropogenic sources and industrial wastes (oil refinery, coal, pharmaceutical, agro-chemical, explosive, Masonite, plastic, etc.), have been accumulated into the environment, resulting fresh and clean water crises [34, 40, 143].

The effects of phenolic wastewater pollutants are huge and unavoidable. They often have accumulated into the body, break salutary food chains resulting in vulnerable environmental consequences. For example, it has been shown that the phenolic water contaminants of >50 ppb are injurious for aquatic species, whereas drinking of 1.0 g phenol could be fatal for human [4]. Moreover, phenolic contaminants have shown to decrease the overall efficiency of wastewater treatment plants [14]. For instance, high concentrations of phenolic water pollutants have inhibited the growth of beneficiary microorganisms, which has been used for treating municipal wastewater in water treatment plants [14].

Figure 2.5 shows some common phenolic water pollutants with their interconversion fates. Herein the catechol, pyrogallol, 3,4-dihydroxyphthalic acid, vainillic acid, 4,5-dihydroxyphthalic acid, cis-4,5-dihydroxy-cyclohexa-1(6), 2-diene-1,4-dicarboxylic acid, 3-hydroxybenzoic acid, 3-dehydroshikimic acid and 4-hydroxybenzoic acid could be converted into the 3,4-DHBA by simple decarboxylation, oxidation, or dehydration reactions.

In addition, 3,4-DHBA and its other phenol-type derivatives which have bactericidal effects could increase their availability in nature [66]. The 3,4-DHBA is commonly found in fruits and vegetables such as olives, grape wines, plant-derived beverages, and so on [71, 84, 142]. Therefore, food processing industrial

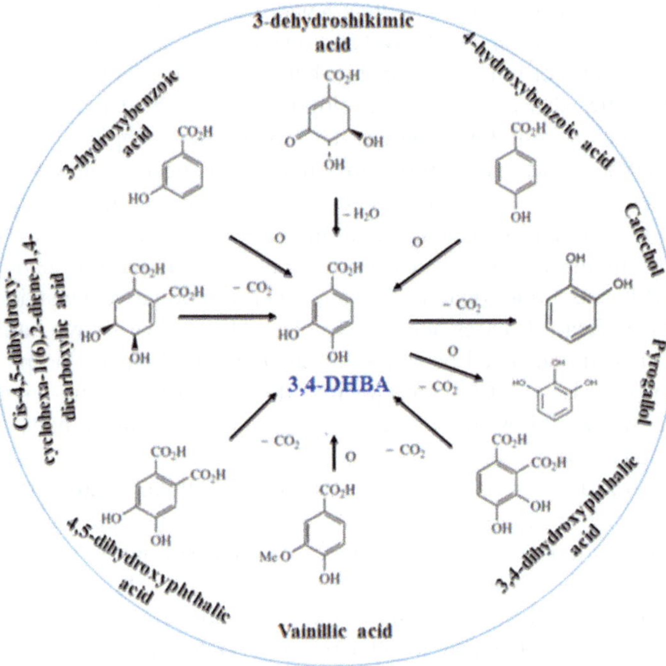

Fig. 2.5 Schematic diagram of the interconversions of some major phenolic water pollutants into the 3,4-DHBA

wastewater effluents have contaminated with a higher concentrations of 3,4-DHBA that continuously dwindling our limited fresh water resources day by day [112]. These have attracted scientists to test the toxicity effects of 3,4-DHBA on both animal, and human tissues and observed contradictory biological effects [10]. Some studies have claimed that 3,4-DHBA actuates as an antioxidant at lower concentrations for the liver, stomach, pancreas, colon, urinary bladder, skin and oral malignant cells [43, 92, 96, 125–128]. But other studies have assured that it induces oxidative stress, hepatotoxicity, and neurotoxicity and promoted tumor formation [10, 93]. However, Babich et al. [10] examined detailed effects of 3,4-DHBA concentrations from 0.5 to 25 mM on normal human cells (S-G and GN61), malignant (HSG1), non-malignant (HSC-2) and carcinoma (CAL27) cells derived from the human oral tissues as shown in Fig. 2.6 [10]. This shows 5, 7.5, and 10 mM of 3,4-DHBA could induce significant initial cytotoxicity ($p < 0.01$) toward S-G, CAL27; and HSG1, HSC-2, GN61 cells, respectively. By observing the similar trends of sensitivities of normal and malignant cells, the group hypothesized that the effects of 3,4-DHBA on malignant cells were dormant. Moreover, it induced toxicity towards the normal human cells, which could be increased with increasing the concentrations of 3,4-DHBA. However, nontoxic level of 3,4-DHBA was observed between 0.5 and 2.5 mM for all cell lines.

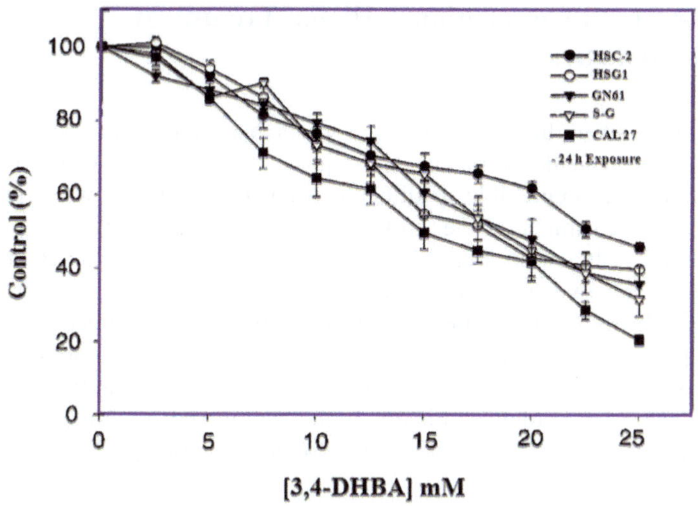

Fig. 2.6 Cytotoxic effects of 3,4-DHBA towards the different human cell lines as determined by neutral red assay. The figure is adapted with permission from Wiley [10]

Although a few studies such as Fenton [33, 112], adsorption [115], O_3/UV or H_2O_2/UV [13] and microbial degradation [17] have been adopted to eradicate 3,4-DHBA from water; they are less sensitive, time consuming, and generating chemical masses to the environment [60]. Therefore, implementation of those methods to tackle 3,4-DHBA is unthinkable, and the paradigm calls to develop novel water purification technology to mitigate 3,4-DHBA from water.

Second, the inorganic pollutant such as arsenic which has been widely recognized as a deadly poison and causative agent of various lethal diseases [5]. This pollutant has been affected 140.0 million people in more than 70 countries [135]. Additionally, other metals such as mercury, lead, and chromium have also shown harmful effects on the living systems [5]. Radioactive metals, e.g., uranium, radium, thorium, etc., have been commonly found in industrial wastewater effluents. These substances are more toxic and lethal even if consumed at lowest level. Other metals such as cadmium, copper, zinc, and cobalt have shown multiple pathological effects such as diarrhea, anemia, nausea, cancer, renal failure, ulcers, gastro-intestinal tract infections, poor growth, fever, lung irritation, and so on [5, 97, 149].

Finally, biological water pollutants mainly bacteria, viruses, fungi, and algal blooms have shown to occur many water born diseases [29, 35]. These pollutants could contaminate beaches that lead to decrease recreational value of water resources with social disturbances. Therefore, effective handling of both the conventional priority and newly emerging pollutants is prerequisite to maintain a healthy ecosystem and disease-free society and nation. Thus water pollution is a chain of reactions, and even a single substrate can affect entire system, hindering social progress and economic growth in daily life processes.

2.3 Threats to Conventional Water Treatment Technologies

Changing of human lifestyles over the years has consistently added different notorious anthropogenic pollutants into aquatic matrices. Eradicating these toxic pollutants is either impossible or difficult using the traditional water purification methods. Here we summarize the most common threats of current water purification technologies as shown in Fig. 2.7. It clearly reflects that a single method is insufficient to remove multiple water pollutants. Most of the methods typically depend on influent water qualities such as turbidity, pH, and temperature that could decrease pollutant sensing ability. In addition, low concentration and small dimensions of some water pollutants are difficult to sense and mitigate by existing water purification technologies. Finally, high operating costs shrink the availability of effective technologies especially in developing countries [131].

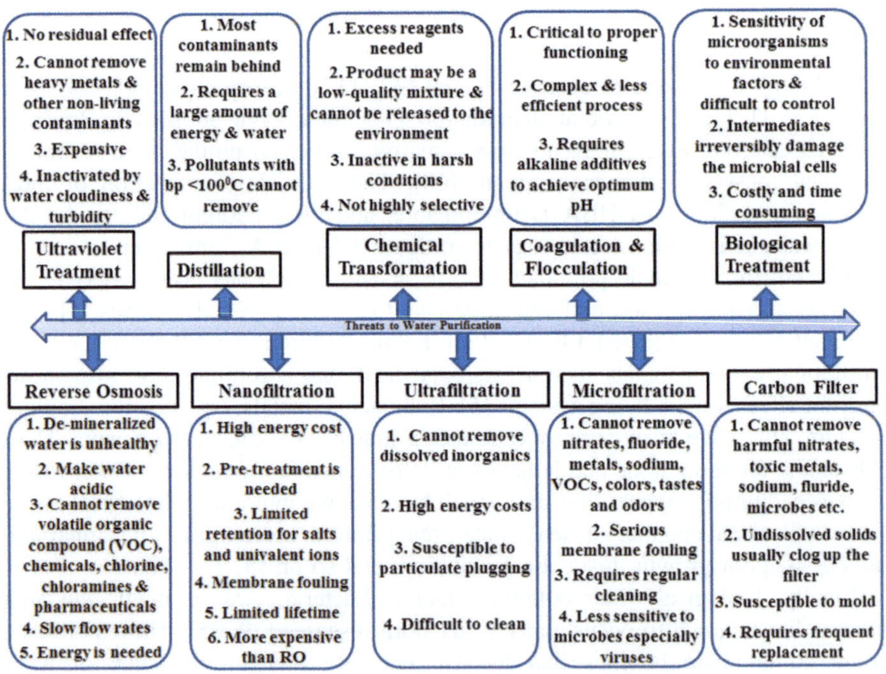

Fig. 2.7 Some major threats of conventional water purification technologies

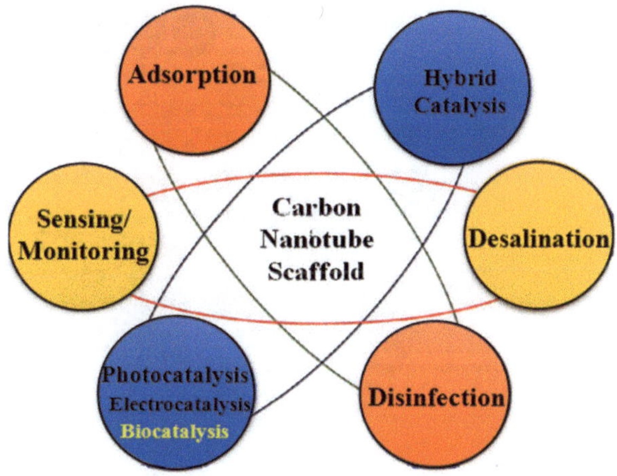

Fig. 2.8 Scaffold functions of CNTs in water treatment technologies

2.4 Carbon Nanotube Scaffold Functions in Water Purifications

Figure 2.8 illustrates the scaffold functions of CNT in water purifications. According to Fig. 2.8, CNT has been widely used as excellent adsorbent media at first for multiple organic, inorganic, and biological water pollutants [6, 107, 131]. Second, CNTs could be used as hybrid catalysts where they may participate in photocatalysis, electrocatalysis, and as nanocarrier for enzyme immobilization or Nanobiohybrid catalysis. These have added new dimension in catalytic degradation of water pollutants. Herein CNTs have made a suitable platform for accelerating the catalytic rates [106]. Third, doped CNTs in various membrane engineering technologies have been used for both the sea and brackish water desalinations [21]. They can be used as pores in engineered membranes or membranes themselves by aligning them together. Finally, high electrical conductivities have made CNT suitable for electrochemical sensor technology to trace and mitigate contaminants, pathogens and high complexity of wastewater matrices [87, 106].

2.4.1 Adsorption

Adsorption is a process where substance (pollutant) concentrations are increased and adsorbed into the layers/sheets of a solid material owing to the operation of surface forces. Basically, it is a surface phenomenon where the substance is adsorbed that is called adsorbate and creates a film on the adsorbent (where adsorbate are adsorbed) surface. Adsorption present in CNTs has been frequently used for capturing wide variety of water pollutants [44, 89, 108, 131, 144]. Some CNT properties have made them versatile for adsorptive processes. For

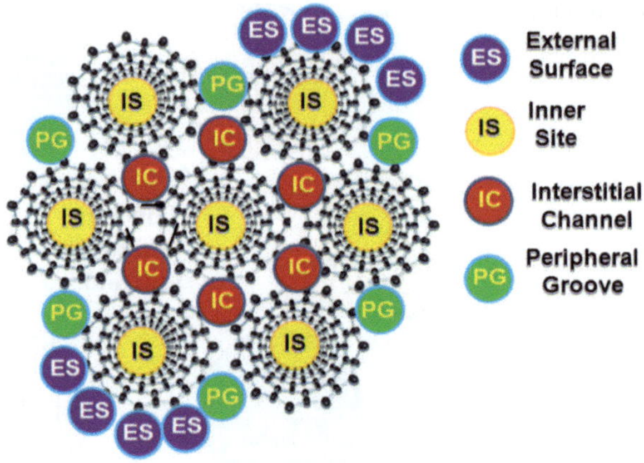

Fig. 2.9 Structural representation of four predicted major adsorption sites of CNTs in a bundle. Close-ended CNTs: adsorption takes place PG > ES > IC, whereas open-ended CNTs: adsorption proceeds IS of open-ended CNT walls > forms 1D chains in the PG > filling of the remaining axial sites of IC > completion of a quasi-hexagonal monolayer on the ES

example, (a) the total CNT surface area is high (SSA 100–300 m^2/g) for high sorption capacity, (b) high pore volume of fibrous material increases surface accessibility and (c) malleable surface charge provides control to select for a specific water pollutant. The effect of CNT morphology and topology on transport and diffusion of water pollutants is displayed in Fig. 2.9.

The external surface, inner site, interstitial channel, and peripheral groove constitute the four possible sites for adsorption (Fig. 2.9). Most of the organic pollutants sorb at external surface and inner site of open-ended CNTs. The external surface of CNTs can be functionalized for attaching both organic and inorganic pollutants. MWCNT usually contributes more pore volumes of inner sites than the SWCNTs. SWCNTs are prone to form bundles because of their strong van der Waals forces along the tube length axis that results in formation of interstitial channels and peripheral grooves (a positive influence for adsorption kinetics). But as grown SSA of SWCNT (400–900 m^2/g) and MWCNT (200–400 m^2/g) which is substantially decreased upon aggregation (a negative influence for adsorption kinetics). SWCNTs of same diameter packed into a bundle (homoaggregation) which results in uniform diameter of interstitial channel. Mismatch diameter in SWCNT bundle (heteroaggregation) poses larger diameter of interstitial channel than homoaggregation. Similar observations have been found in aggregated MWCNTs. Interstitial channels have found important for sorbing PAHs (e.g., naphthalene and tetracene) water pollutants displaying diameter <1 nm. It has been shown that the diameter of intertubular spacing of the bundle structure of 1.43 nm, which increased from 1.43 to 2.20 nm upon tetracene adsorption into SWCNT. Sorbate configuration determines their interactions with peripheral grooves area of aggregated CNTs. Inorganic molecules adsorb at peripheral groove sites, and the rate might be faster in close-ended CNTs. The site is generally reached to equilibrium faster than the inner sites because of its external face to the solvent.

Functionalizing CNT can mediate specific pollutant adsorption, increases CNT colloidal stability and chemical reactivity. Three different methods have been explored for CNT functionalizations such as (a) physical non-covalent wrapping, (b) covalent wet chemical agent treatments, and (c) endohedral filling of CNT hollow cavity. Major interaction forces between CNT functionalities and water pollutants are covalent bonding, hydrogen bonding, electrostatic interactions, ion exchange, hydrophobic interactions, π–π electron coupling, and mesopore filling. Some important evidences of the roles of CNT functionalities for pollutant adsorptions are illustrated in Fig. 2.10. It ensures both the organic and inorganic wastewater pollutants can effectively bind with the functionalized CNT. Although CNT functionalization could substantially increase sorption of various water pollutants, it will decrease CNT ability for binding of hydrophobic organic water pollutants, e.g., naphthalene, phenanthrene, pyrene, and polychlorinated biphenyls that interact strongly with virgin CNT through π–π and hydrophobic interactions. Chemical functionalizations disrupts π–π carbon networks and increase wettability of surface. Therefore, great care needs to be taken to functionalize a material in a controlled fashion, so that one can preserve hydrophobic CNT skeleton and immobilize hydrophilic groups for multi-pollutants wastewater treatment.

After absorbing the pollutants for a certain concentration into the adsorbent, equilibrium is reached and the relationship between the amounts of pollutant adsorbed and in water, is called an adsorption isotherm. The kinetics of this mechanism have been evaluated mathematically using different models such as Langmuir, Freundlich, Halsey, Henderson, Smith, Elovich liquid film diffusion, intraparticle diffusion, and Lagergren [5]. Wang et al. [133] observed the adsorption of Pb(II), Ag(I), Cu(II), and Co(II) ions on MWCNT, and the adsorption data were fitted well to the Langmuir model as shown in Fig. 2.11 [133].

Figure 2.12 shows some observed evidences of pollutants adsorption onto CNT surfaces. As we can see, TEM image of MWCNTs after Pb (II) adsorption which is not uniform and mainly adsorb at the tips and defective sites of the MWCNTs (Fig. 2.12b) [133]. SEM image of *Eschericia coli* (*E.coli*) bacteria exposed to SWCNTs clearly suggests loss of their morphology (Fig. 2.12c) [53]. Sponge-like CNT adsorbent floated on oil polluted water and simultaneously can remove oil with a large adsorption capacity (from 80 to 180 times their own weight for a wide range of solvents and oils) (Fig. 2.12d) [38], and finally salt adsorption by CNT membranes (Fig. 2.12e) [141].

2.4.2 Hybrid Catalysis

CNT has been popular in photocatalysis of multiple water pollutants because of their high electrical conductivity and charge transferring ability [26]. Good electrical conductivity of CNT might be resulted from the presence of reactive surface

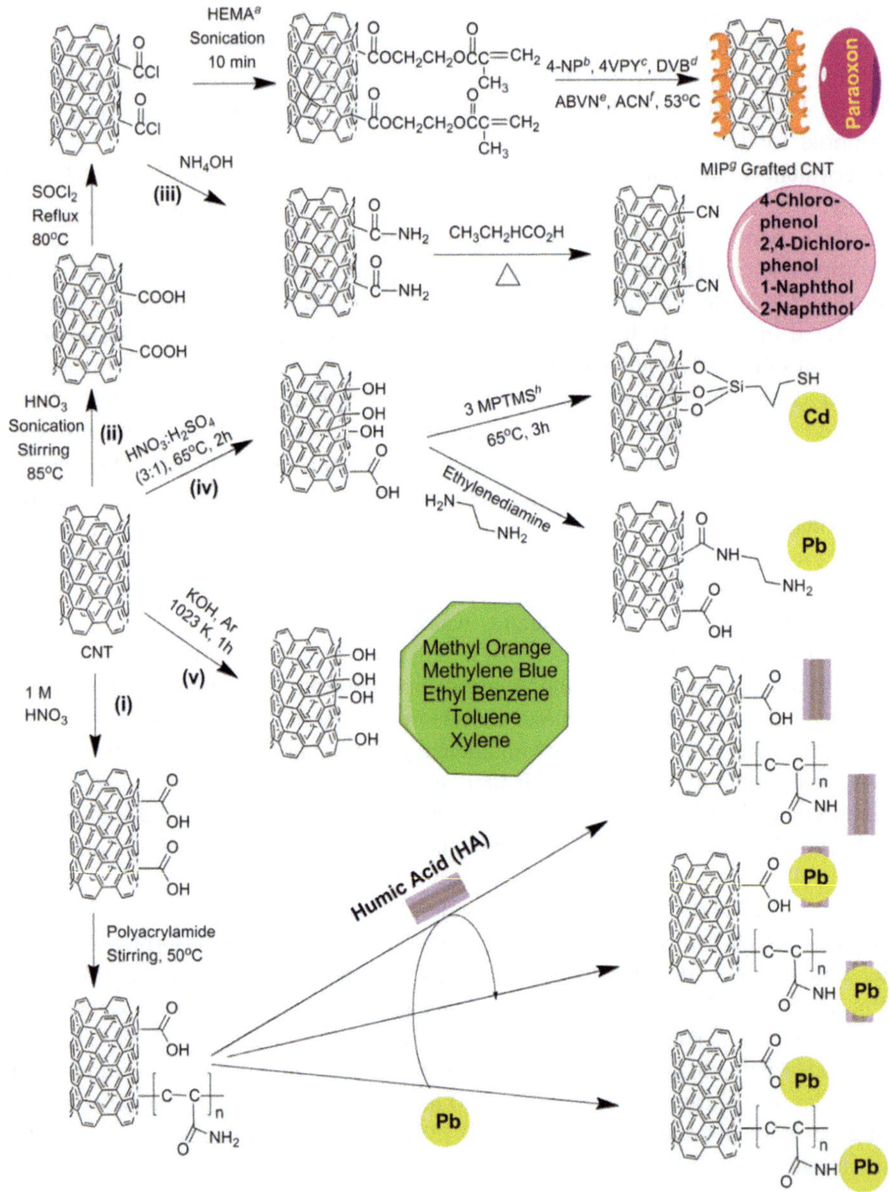

Fig. 2.10 Functionalization of CNT as adsorbent for organic and inorganic water pollutants. Here [a]2-hydroxyethyl methacrylate, [b]4-nitrophenol, [c]4-vinyl pyridine, [d]divinylbenzen, [e]2,2-azobis (2,4-dimethyl)valeronitrile, [f]acetonitrile, [g]molecularly imprinted polymers and [h]3-mercaptopropyltrimethoxysilane

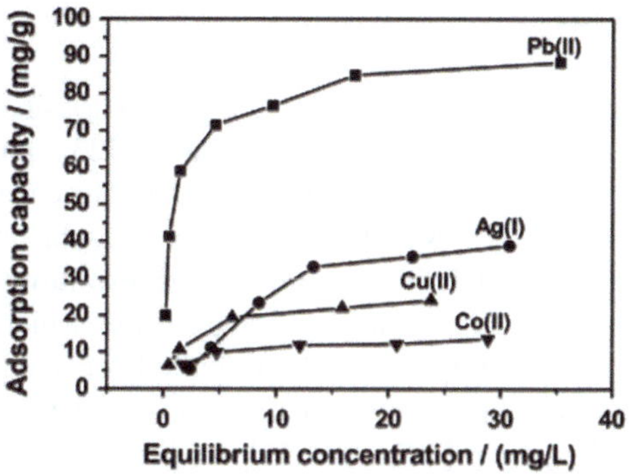

Fig. 2.11 Adsorption isotherms of acidified MWCNTs to Pb (II), Ag (I), Cu (II), and Co (II) at 25 °C. Figure is adapted with permission from Elsevier [133]

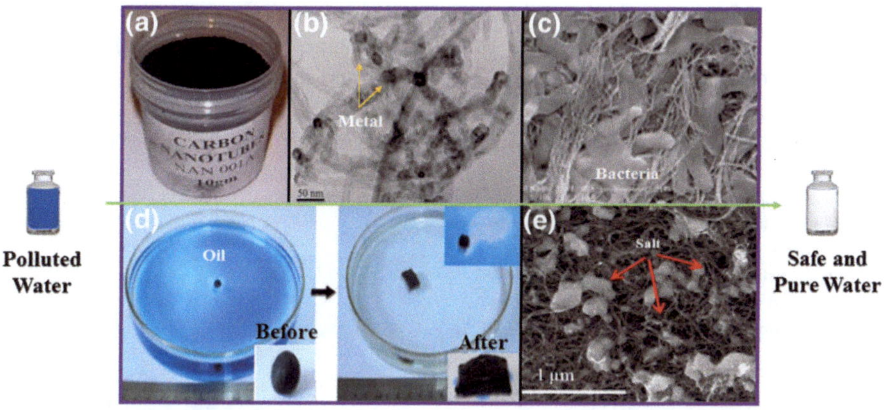

Fig. 2.12 Adsorption of water pollutants on CNT surfaces. **a** Pristine CNTs, **b** metal, **c** bacteria, **d** oil and **e** salt adsorptions onto CNT surfaces. The figures are adapted with permissions from Elsevier, ACS, Wiley and Sons and Nature [38, 53, 133, 141]

groups, sp^2 carbons, curvature pattern, topological defects, and edge-plane sites [111]. Another potential hybrid catalyst has been used to degrade water pollutants in a process called catalytic wet air oxidation (CWAO). This can be defined as the dissolved and suspended water pollutant's oxidation using an oxidizing agent such as O_2 from air in the presence of catalyst. In addition, recently CNTs have also been used for immobilizing enzymes [28]. Immobilized enzymes have shown good thermostability, broad pH ranges, and higher reusability in the fields of water purification [6, 90, 137], pharmaceuticals, chemical synthesis, biosensors, and so on [62]. The suitability of CNTs for enzyme attachment is because of their known atomic structure, big surface area, mass transfer, and good enzyme loading capacities which help to develop an effective Nanobiohybrid catalyst for water purifications [137].

2.4.2.1 Photocatalysis

CNTs have been used as reinforced photocatalytic composite materials along with other semiconductors such as TiO_2, ZnO, and so on [23]. Photocatalysis means a photoreaction that is accelerated by light/photon in the presence of single or multiple catalyst(s). Functionalities on CNTs have shown important interactions for doping CNTs with different photocatalytic agents [113]. These have increased CNT total surface area, defects, electrical conductivity and so on, which enhance the overall photocatalytic rate. Two methods have been applied for synthesizing CNTs based photocatalytic composites. First, covalent and non-covalent attachment of desired nanoparticles on CNT surface [102]. Second, in situ synthesis of nanoparticles of interest followed by electrochemical deposition [140]. The main advantage of this method is to get a uniform composite nanomaterial produced by selective deposition of nanoparticles in the CNT active sites. Therefore, CNT can be considered as suitable support for multiple catalysts doping in photocatalytic reactions.

Figure 2.13 shows the possible mechanism of actions following CNTs for enhancing photocatalytic activity. Band gap or energy gap is a common phenomenon in solid state physics, and can be defined as energy intervals (ΔE_g) (no electrons exist) between the valence and conduction bands (V/CBs). VB is the highest energy state with electrons, whereas CB is the lowest energy band in the

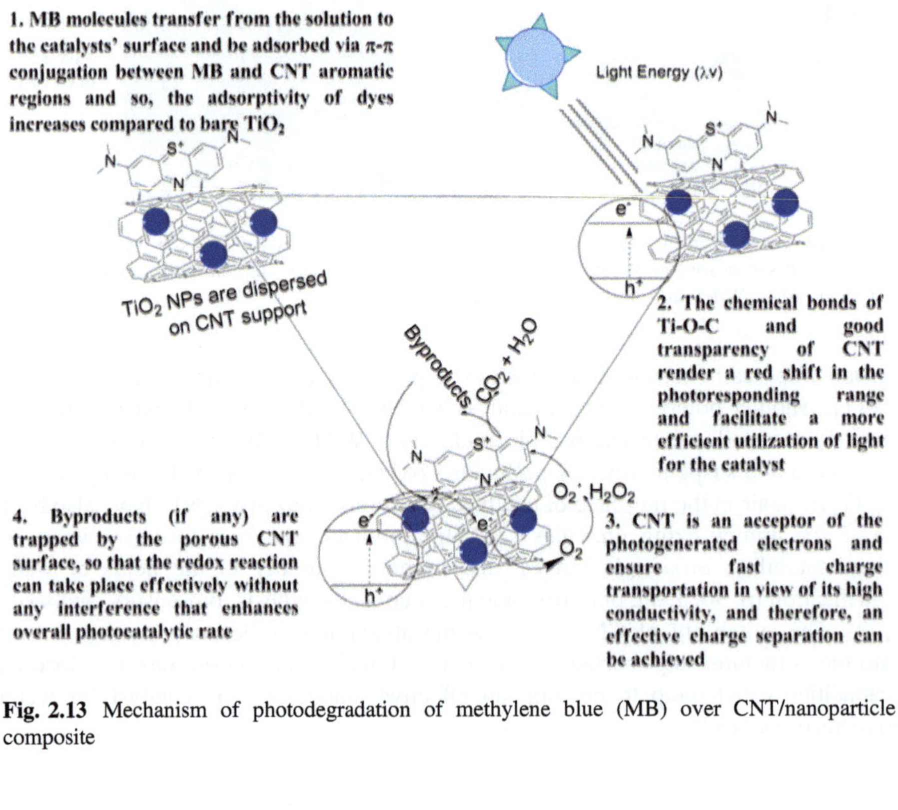

Fig. 2.13 Mechanism of photodegradation of methylene blue (MB) over CNT/nanoparticle composite

absence of electrons. Photons from different light sources are exposed to a catalyst nanoparticle, the VB electrons getting excited then and rapidly move to the CB. This creates a vacancy or hole (h^+) in VB. These holes (h^+) then react with H_2O or OH^- ions followed by hydroxyl radical ($^{\bullet}OH$) formation that directly oxidizes the adsorbed water pollutants on CNT surfaces. On the other hand, the excited electrons from VB to CB form $^{\bullet}OH$ which reacts with O_2 followed by the formation of superoxide radical ion ($O_2^{-\bullet}$). This highly reactive radical attacks and oxidizes the target pollutant rapidly. However, the major drawback in photocatalysis is to have a chance of recombination of charges avoiding photodegradation. This electron–hole recombination can be prevented by making a composite CNT material with other semiconductor nanoparticles. CNTs in metal composites act as insulating nanomaterial or barrier junction, which inhibits the recombination and increases photocatalysis rate. CNT can capture electrons due to its high charge conducting ability that ultimately gives less chance to electrons for recombining with holes. Consequently, CNTs could be sponsored as photo-generated electron acceptor that enhances interfacial electron transfer process, whereas the semiconductors are good electron donors under irradiation [32, 109, 114, 122]. But, the overall process has also been regulated by some factors such as the nature and type of semiconductors, light radiation, pH, temperature, and target pollutant concentrations [64, 65].

2.4.2.2 Catalytic Wet Air Oxidation

CWAO has considered one of the most effective water treatment methods, and has commercially been used for last 60 years [78]. CNTs have been used along with other catalysts such as Pt, Pd, Ru, and so on [31, 139] to increase catalytic efficiency and decrease costs [78]. Yang et al. [139] reported MWCNTs for CWAO of phenol in a batch reactor as shown in Fig. 2.14 [139]. Carboxylic groups (–COOH) on CNTs showed good catalytic activity in phenol oxidation (1000 mg/L) at 160 °C and 2.0 MPa (Fig. 2.14b). Liquid O_2 adsorbed, and dissociated on MWCNTs graphite sheets produced dissociated oxygen atom (DOA) [88]. Functionalities mainly –COOH groups generated by treating MWCNTs with H_2SO_4/HNO_3 and the DOA could form HO_2. Then the radical could mineralize phenol to CO_2, H_2O and low organic compounds [111]. Kim and ihm [61] summarized the degradation of major industrial organic water pollutants by CNTs based CWAO process [61].

2.4.2.3 Nanobiohybrid Catalysis

Figure 2.15 illustrates a Nanobiohybrid catalyst prototype for sensing, monitoring, and degrading of multiple pollutants in water. The Nanobiohybrid could be made by attaching enzymes on CNTs via one of the three major routes such as binding to a support (physical adsorption and covalent bonding), cross-linking (carrier free) and encapsulation or entrapment [48]. In physical adsorption, spontaneous

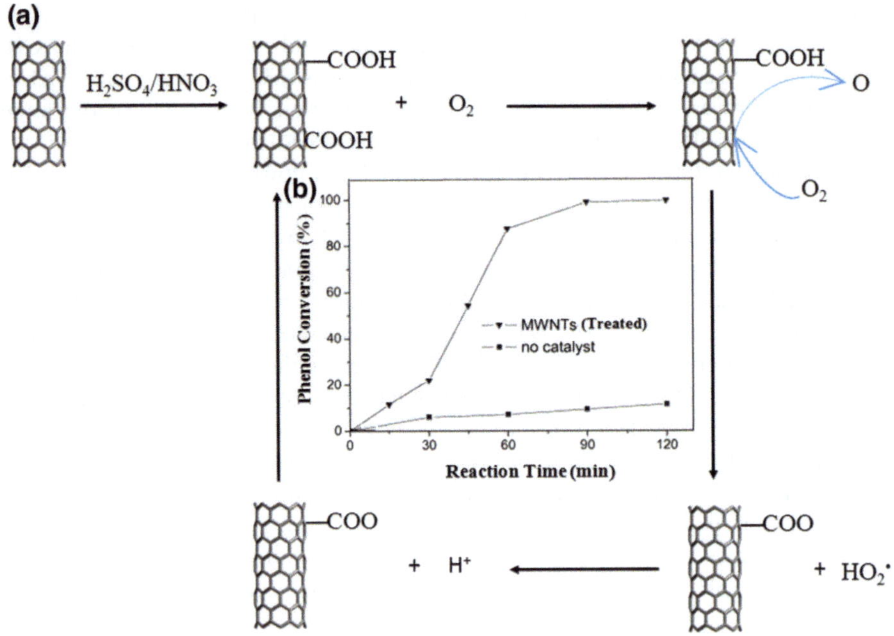

Fig. 2.14 **a** Mechanism of producing the radical (HO_2^{o}) in CWAO for phenol degradation by H_2SO_4/HNO_3 treated MWCNTs and **b** efficiency of phenol removal. The figure is adapted with permission from Elsevier [139]

adsorption of enzymes into CNTs has occurred due to the inheritance hydrophobic and electrostatic interactions between CNTs and enzyme catalyst (Fig. 2.15I) [30, 95].

Covalent bonding of enzymes with CNT has been popular by inducing the reaction of the free amine groups on the surface of enzymes with –COOH groups that could be generated by sidewall oxidation of CNTs and subsequent activation using carbodiimide chemistry (Fig. 2.15II) [9, 47, 51]. Cross-linking polymers such as chitosan, poly(diallyldimethylammonium chloride), and so on can also be used to immobilize enzymes on CNTs [69, 123]. In addition, LBL approach has been adopted for immobilizing enzymes using enzymes encapsulation process. It permits the coating of various enzymes, producing multilayer enzyme films on CNTs [28] as shown in Fig. 2.15III. The covalently bonded enzymes would be better for the Nanobiohybrid applications in water purifications.

This is due to least chances of enzyme leaching from the supports compared with non-covalent approach [28]. Non-covalent enzyme immobilization is typically popular for fabricating the biosensor devices. This is because of maintaining intact CNT skeletal integrity and enzyme configurations that yield higher electrical conductivity [28]. However, encapsulation or entrapment methods have been popular for attaching more than one enzyme molecule. This method can be used to develop

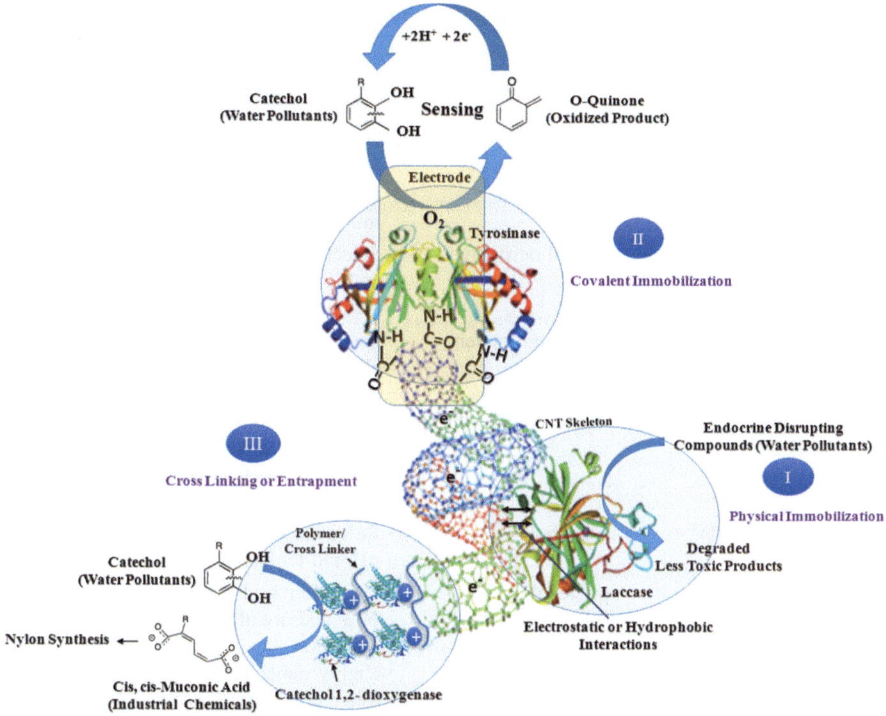

Fig. 2.15 Mechanisms of a potential Nanobiohybrid catalyst for water purification

a single biohybrid catalyst which can degrade a range of different water pollutants by multiple immobilized enzyme complexes.

Here we summarized some major observations of Nanobiohybrid-mediated water purifications in Table 2.1. However, besides CNTs, there are many other supports such as glass beads [85], agar [101], hydrophobic, and ionic supports [103], membranes [80], nanofibers [137], nanoparticles [82] and the combination of different supports [91] have been well documented for the various enzymes immobilizations which have also shown higher thermostability, shelf life, and reusability and successfully used for multiple applications.

Nanobiohybrid has some advantages over conventional chemical oxidation of water pollutants [8, 15, 28]. It can be applied for water purifications where other chemical transformations of water pollutants are not possible. In addition, the catalyst has greatest efficiency for disintegrating the electron resonance of benzene ring containing aromatic water pollutants in contrast to photocatalysis. Furthermore, recovered Nanobiohybrid catalyst could be used for multiple times [123], so the method is economically viable. Consequently, the hybrid can play three major functions with high selectivity and sensitivity in water purifications (Fig. 2.15):

Table 2.1 Some major evidences of the actions of Nanobiohybrid catalysts in water purifications

Material type	Water pollutant	Major observation	Refs.
MWCNTs-Laccase	ABTS[a], Bisphenol and Catechol	• Enzyme loading capacity was highest for O-MWCNTs and lowest for C_{60} • No obvious structural change of enzyme observed after immobilization	[100]
MWCNTs-Tyrosinase	Phenol derivatives	• Selectively oxidized phenol and its derivatives with high stability of the catalyst	[123]
SWCNTs-HRP[b] SWCNT-DM[c]-Con A[d]-HRP	Dibenzothiophene (DBT)	• Increased enzyme loading with high specific activity triggered efficient DBT degradation • DBT removal rate: 66.46 and 94.65% for SWCNT-HRP and SWCNT-DM-Con A-HRP catalysts, respectively	[73]
SWCNTs-1,2 HQD[e]	Catechol	• Removed catechol with wide pH and temperature ranges	[124]
MWCNTs-SBP[f]	p-Cresol	• Nanobiohybrid retained 40–60% activity after multiple time uses • Highly stable and reusable than free enzyme in removing p-Cresol	[9]
SWCNT, MWCNT, SWCNTs-Crude enzyme, MWCNTs-Crude enzyme	Aniline	• Enzymes adsorption increased in MWCNTs than SWCNTs • Effective aniline removal observed in SWCNTs-enzyme compared with MWCNTs-enzyme indicated efficient reaction between aniline and enzymes on the surface of SWCNTs. This played a key role in the rapid enzymatic biodegradation of aniline	[138]

[a]2,2′-azino-bis-(3-ethylbenzthiazoline-6-sulfonic acid) diammonium salt, [b]horseradish peroxidase, [c]n-dodecylβ-D-maltoside, [d]Concanavalin A, [e]hydroxyquinol 1,2 dioxygenase and [f]Soybean peroxidase

(i) binding and pre-concentrating the pollutants; (ii) removal and/or degradation of pollutants; and (iii) sensing and monitoring the pollutants in water. Moreover, it can be used in industry for yielding commercially important compounds or value-added chemicals which have been produced by mitigating of organic wastewater pollutants (Fig. 2.15III). In short, the technique would have double benefits—it reduces total water contaminants, and simultaneously this increases commercial value.

2.4.3 Desalination

On the basis of current fabrication systems, there are two types of CNT membranes such as (i) vertically aligned (VA) and (ii) mixed matrix (MM) CNT membranes [3] as shown in Fig. 2.16. The VACNT membranes can be synthesized by aligning perpendicular CNTs with supportive filler contents (epoxy, silicon nitride, etc.) between the tubes (Fig. 2.16a) [42]. On the other hand, a MMCNTs membrane consists of several layers of polymers or other composite materials (Fig. 2.16b). These membranes work with low energy consumption because of CNT's frictionless water transport capability through nanotubes hydrophobic hollow cavity. The membrane is highly sensitive toward the multiple water pollutants and salts. In addition, due to the CNT cytotoxicity, the CNT membrane has antifouling and self-cleaning abilities with high recrudescence and reusability facilities. For instance, Dumée et al. [24] observed the highest water permeability of SWCNTs (5,5), and (8,8); and sodium ion retention was reached (99%) in (5,5) and (6,6) SWCNTs at flux rates of 12 kg m^{-2} h^{-1} [24]. Moreover, functionalities such as positive (–NH_3^+), negative (-COO^-, sulfonic acids) and hydrophobic (aromatic) groups could be implanted by using different wet oxidizing agent treatments of

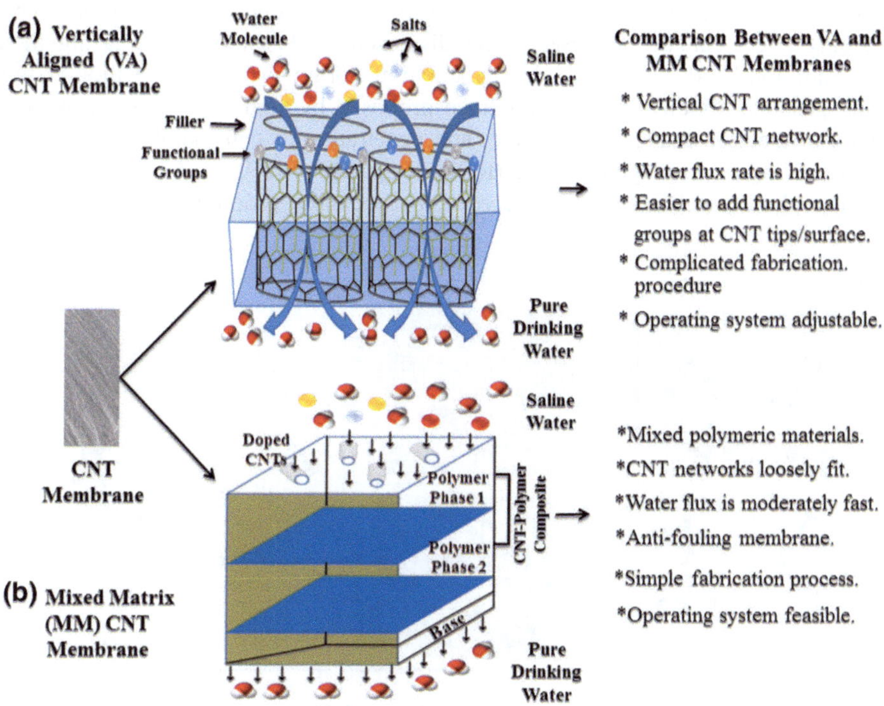

Fig. 2.16 Schematic illustration of two types of CNT membranes. **a** VA and **b** MM–CNT membranes

CNT [36, 56]. The groups can reject a selective water pollutant and also attract water molecules through nanotube hole. Chan et al. [19] simulated and fabricated CNT membrane (diameter 1.5 nm) with two zwitterions at tip ends. The group was achieved 100% ion rejection, which was 0.0% in nonfunctionalized membrane [19].

2.4.4 Disinfection

Aggregated mesoporous CNTs [41] have been utilized to remove biological contaminants such as bacteria and viruses from water. Figure 2.17 shows some possible mechanisms for disinfecting microbes adsorbed onto CNT sheets.

Cytotoxic CNTs have shown higher pathogens decontamination rate [54] compared with AC [18, 119]. Therefore, pathogen contaminated effluents could be found in AC treated water. Liu et al. [76] studied the toxic effects of pristine SWCNTs on both Gram-positive bacteria such as *Staphylococcus aureus* (*S. aureus*), and *Bacillus subtilis* (*B. subtilis*) and Gram-negative bacteria such as *E. coli* and *Pseudomonas aeruginosa* (*P. aeruginosa*) [76]. The group observed that the dispersed individual CNT could puncture cell membrane integrity rapidly and strongly than aggregated CNTs. Soft and smooth cells such as Gram-positive bacteria could be more vulnerable to attack by CNTs than Gram-negative bacteria. This effect can be vice versa depending on the cell membrane compositions. This piercing effect can be increased using dispersed CNT solution, increasing their concentrations and shaker speed augmentation during incubation. Although few other possible toxicity mechanisms have proposed by some groups such as inhibition of electron transports, leakage, and penetration of cell membrane and generation of reactive oxygen surface (ROS) (Fig. 2.17b) [39, 52, 63, 72, 94, 104]; most of these mechanisms are not yet experimentally proofed. Kang et al. [54]

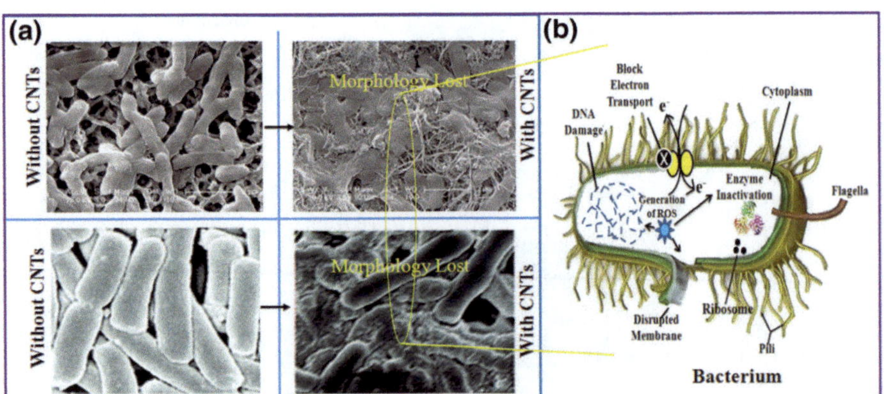

Fig. 2.17 **a** SEM images of bacteria before and after CNT exposures, and **b** some possible mechanisms of CNTs mediated bacterial cell death. Figure **a** is adapted with permission from ACS [52, 77]

observed the dominant toxicity resulted from using SWCNTs than MWCNTs [54]. The group oxidized 4.1 μm of MWCNTs by H_2SO_4/HNO_3 that yielded higher toxicity than 77 μm of MWCNTs bundles in diameters. It clarifies size-dependent toxicity mechanisms and can be a key parameter to ascertain antibacterial activity. Kang et al. [54] hypothesized higher toxicity induced by short, unzipped, and dispersed MWCNTs toward bacteria [54]. Other factors such as incubation time, buffer types, concentration, and functionalities have shown regulatory effects on CNT antibacterial activity [7, 55]. As an initial foray into the CNT microbial cytotoxicity, we confirmed that the scientists, who have been active in the field, only focused on bacteria physiological changes upon CNT treatments using SEM, AFM, and TEM. However, only these characterization methods cannot accurately explain the effects of CNT toxicity on bacterial cell proliferation. More biological experiments such as pattern of gene expressions, transcriptomics, proteomics, and genomics data comparisons along with other system biology approaches of CNTs treated bacteria could be effective for settling the mechanisms of cellular deaths. In addition, misinterpretation could be resulted from the use of pristine CNTs contaminated with amorphous carbons, metal catalysts, nanofibers, etc. Liu et al. [77] hypothesized the CNT toxicity can be dependent on its geometrical compositions and surface functionalities [77]. Therefore, more study is necessary to acquire the nature of bacterial cell death when CNT is an adsorbent media and should be treated with cautions.

2.4.5 Sensing and Monitoring

Rapid sensing of trace water pollutants has remained a critical job by using conventional sensor technologies. Moreover, complex wastewater matrices have increased this difficulty level. The paradigm has called to develop highly sensitive and fast responsive CNT-based devices or sensors. This is because of some CNT specialties such as fast electron conductivity, high adsorbing power (so that water pollutants can concentrate on CNT surfaces before sensing), hydrophobicity, high recovery rate, and increased reaction rate. As shown in Fig. 2.15II, enzymes immobilized on CNTs oxidize the water pollutants followed by electrons transformation for high signal detection. In addition, charged or ionic pollutants adsorbed into CNTs have given electrical conductance by measuring the ratio between analyte concentrations and current fluctuations [86]. López and Merkoçi [79] designed enzyme (tyrosinase)-integrated-CNT-epoxy composite electrode (CNTECE-Tyr) and compared with tyrosinase biosensor based on a graphite epoxy composite (GECE-Tyr) for quantifying phenolic water pollutants such as catechol [79]. Herein CNTECE-Tyr electrode showed higher signals reflecting the good electrocatalytic properties of CNTs over graphite sheet. It was of 46 and 294 mA/mM cm^2 for GECE-Tyr and CNTECE-Tyr, respectively with a current intensity 90% higher in CNTECE-Tyr compared with GECE-Tyr.

2.4.6 Research Gaps and Challenges

Pristine CNTs are often contaminated with various metal catalysts, ash and carbonaceous agents. These act as additional adsorbent sites of CNTs towards the water pollutants. The impurities have determined nanotube's pore diameters, morphology and could influence or inhibit adsorption behaviors [2]. In catalysis, metal and amorphous carbon impurities have contributed towards the observed catalytic behaviors and mislead total electrocatalytic activity [12]. Furthermore, CNT impurities could have several detrimental effects on enzyme immobilizations. First, non-nanotube carbon materials could inhibit maximum immobilization efficiency. Second, metals in pristine CNTs could thwart immobilized enzyme's reaction rates. Moreover, different CNT geometries and impurities might complicate the procedures and compromise CNT performances in sensor devices. Separating metallic from semiconducting CNTs has remained a challenging job because of their complex carbon networks. Synthesis of homogenous CNTs with perfect geometry for high power density is also appreciable for upgrading this technology. One pot combination of CNTs with classical electrochemical techniques could be a nice electrical device for water pollution control. It would increase the overall catalysis rate, fast mass transport, good control over the classical current flow in small geometry.

Although several approaches such as gas and liquid phase reactions [74], direct oxidation [70, 147], electrochemical oxidation [27], and wet chemical purifications [22, 98, 134] have been adopted to remove CNT impurities and manipulate solubilities, they often degrade CNT, shortening, and burning under extreme heat and excessive chemicals. Removing impurities with intact nanotube integrity is one of the toughest tasks in CNT-based water purification applications. For instance, HNO_3 mediated CNT purification could produce defected and curled CNT fragments [22, 75]. Moreover, the method could not remove 100% impurities from CNTs even at elevated temperatures [105]. In order to get intact CNT, several physical methods have been designed such as filtration [117], high temperature annealing [68], and repetitive centrifugations [46]. But the methods could not purify CNTs completely and also have less dispersed CNT samples. Therefore, more studies are necessary to develop novel CNT purification technology which would not only remove CNT generic impurities, but maintain an intact CNT skeleton. The treatment procedure should also anchor appropriate functionalities that might give appropriate overhead to attach other molecules or compounds, needed for upgrading the current water purification technologies.

Defects in CNTs may help in adsorption processes [148], but cannot be suitable in sensor and photocatalytic water purification technologies. Lower electrical conductivity has observed in defected MWCNTs [99] because it scatters the electron transportation of CNTs [81]. However, localized edge-plane defects at CNT tip ends and around the tube walls where one of the concentric tubes terminates could generate high peak current [11]. This observation is also similar with another study [16]. The group observed high charge density at pentagonal defects of nanotube mouths than those observed in the regions of hexagonal graphite. It indicates a relationship between topological defects and CNTs electroactivity [16], which are necessary for developing an effective and highly conductive CNTs based sensor and photo-electronic water purification devices.

Besides CNT purifications, another major hurdle is to manipulate their solubility in water. If anyone wants to remove pollutants out of water, it should be considered to bring wastewater into the reactive surfaces of CNTs. Pristine CNTs are not soluble in water because of hydrophobic graphite sheet [118]. As a result, various covalent and non-covalent functionalization strategies have been adopted [45, 146]. But most of the covalent wet oxidation techniques have failed to give durable CNT solubility in water. Two methods such H_2SO_4/HNO_3 and $KMnO_4$ have been effective for adding –COOH groups and ultimately increased CNT solubility [121]. But higher CNTs solubility might come from H_2SO_4/HNO_3 treated CNTs. Although the method has frequently been used for immobilizing enzymes on CNTs [30], this might not be suitable for sensor and photocatalytic CNT applications. This is because of yielding defected CNTs that might hamper electron transportations. Based on CNT solubility profiles, Upadhyayula et al. [131] hypothesized less soluble CNTs should be used in adsorbing media because of their handling and practical difficulties [131], whereas highly soluble CNTs could be used in composite membrane technology. Furthermore, bacteria have shown highest affinity toward semi-soluble CNTs [131]. Therefore, selection of appropriate wet oxidizing agents for purifying and solubilizing CNTs depends on desired water purification technology where CNT solubility and defect manipulations are required.

In desalination technology, the major challenges are to obtain an effective CNT membrane, controlled CNT growth method, processing and applications. Kar et al. [57] observed difficulties to install a dense membrane having CNTs of 12–13 order of magnitudes/cm^2 [57]. Any irregularities in membrane shape could compromise with water passage and pollutant retention. Other hurdles in desalination are CNT tip functionalization and CNT selectivity to specific pollutant. Highly reactive carbon atoms at CNT mouth could be oxidized with different agent treatments which have been used for CNT purification and functionalization. But such treatments are often corrosive enough that cut CNTs into short fragments leading to membrane leakage. In addition, hydrophilic functionalities generated at CNT tips stimulate fast water transport through the hollow tubes, but have created steric blockage by the ions that are attracted and saturated at CNT mouths [20]. The affinity between functional groups and water molecules may cause a temporary pause of water molecules transportation around the CNT functionalized tip ends. Thus the functionalization is a rate limiting step which must be carefully optimized to tune the CNT membrane permeability to water molecules and solutes. Finally, doping of CNTs into RO, UF, NF, or MF membranes is still a challenging task for upscaling existing water purification technologies. Probably it can be a judicious choice to use costly SWCNTs in small amounts for doping rather than using alone as aligned membrane. However, standard experimental procedure with pilot projects is necessary to integrate CNTs into the existing membranes, which would upgrade overall water purification technologies efficiency with reduced efforts and time [116].

Special attention to CNT toxicity is important because of its various environmental fates and toxicity phenomena. Pristine CNTs have shown toxicity to living organisms [50, 67]. Smart et al. [120] hypothesized three potential properties of CNTs in determining their toxicity levels in the environment [120]. First, high aspect ratio of CNTs helped them to react with cellular membranes followed by their adsorption and transportation of toxic substances into the living organisms. Second, CNT longer retention time made its smooth contact with cell wall that increases the chance for damaging surrounding tissues. Finally, pristine CNT toxic impurities could have greater reactivity and toxicity within the particle. For instance, CNTs with metal ions such as Fe(II), Ni, and so on have shown toxicity to living cells [67].

Although CNT-based water purification technologies have faced some challenges, most of them are temporary. But the feasibility of most of the water purification technologies, such as adsorption, hybrid catalysis, disinfection, sensing and monitoring and so on at large scale yet appeared to be low. The commercial availability of these technologies should be encouraged because these meet most of the commercial features such as high water permeability, desalination capacity, pollutants selectivity, robustness, antifouling, energy savings, material costs, scalability and compatibility with industrial settings.

2.5 Conclusions

Global challenges particularly industrialization, urbanization, and anthropogenic activities have been continuously polluted water, raising the issues of various diseases, food safety and biodiversity of terrestrial, aquatic, and aerial, flora and fauna. In order to overcome the fresh water availability, CNT-based water purifications sowed the seeds of novel and innovative wastewater purification technologies because of its large surface area, high aspect ratio, greater chemical reactivity, lower cost, and energy, less chemical mass, and impact on the environment. Here we reviewed CNT-mediated effective decontamination processes such as adsorption, hybrid catalysis, desalination, disinfection, sensing and monitoring of three major classes such as organic, inorganic, and biological water pollutants. Nanobiohybrid catalysis has recently been grown as a novel technology with high selectivity, sensitivity, stability, and reusability. But a few studies have been adopted and/or published on Nanobiohybrid decontaminations, and most of them have dealt only to determine permissive and non-permissive levels of water pollutant concentrations. It appeals more study to mature the field from various angles not only to sense the novel toxic water pollutants, but to mitigate them from complex wastewater matrices. As a corollary, we emphasized the importance of Nanobiohybrid-mediated water purification technology. We believe that upgrading the Nanobiohybrid technology by controlling its steering speeds and improving its performances would be a robust solution to face current threats, and challenges of water purifications through both incremental and revolutionary ways. Finally, we forayed into the deeper thoughts and compiled promises, facts and challenges of the important water purification technologies. Since water purification is a complex process, hydrologists, membrane technologists, environmentalists and industrialists can design "ONE POT" combination where effective water purification technologies would instate to tackle both the conventional and newly emerging toxic pollutants effectively. With the hope of this, let us think a best future of improved and efficacious decontamination platform, and thereby ensure plethora of fresh water for all.

References and Future Readings

1. ABC-Science: 10 facts about our amazing oceans. http://www.abc.net.au/science/articles/2014/06/04/4018335.htm (2014). Accessed 25 Sept 2014
2. Agnihotri, S., Mota, J.P., Rostam-Abadi, M., Rood, M.J.: Adsorption site analysis of impurity embedded single-walled carbon nanotube bundles. Carbon **44**(12), 2376–2383 (2006)
3. Ahn, C.H., Baek, Y., Lee, C., Kim, S.O., Kim, S., Lee, S., Kim, S.-H., Bae, S.S., Park, J., Yoon, J.: Carbon nanotube-based membranes: fabrication and application to desalination. J. Ind. Eng. Chem. **18**(5), 1551–1559 (2012)
4. Aitken, M.D.: Waste treatment applications of enzymes: opportunities and obstacles. Chem. Eng. J. **52**(2), B49–B58 (1993)
5. Ali, I.: New generation adsorbents for water treatment. Chem. Rev. **112**(10), 5073–5091 (2012)
6. Ali, M., Das, R., Maamor, A., Hamid, S.B.A.: Multifunctional Carbon Nanotubes (CNTs): a new dimension in environmental remediation. Adv. Mater. Res. **832**, 328–332 (2014)
7. Arias, L.R., Yang, L.: Inactivation of bacterial pathogens by carbon nanotubes in suspensions. Langmuir **25**(5), 3003–3012 (2009)
8. Asuri, P., Bale, S.S., Karajanagi, S.S., Kane, R.S.: The protein–nanomaterial interface. Curr. Opin. Biotechnol. **17**(6), 562–568 (2006)
9. Asuri, P., Karajanagi, S.S., Sellitto, E., Kim, D.Y., Kane, R.S., Dordick, J.S.: Water-soluble carbon nanotube-enzyme conjugates as functional biocatalytic formulations. Biotechnol. Bioeng. **95**(5), 804–811 (2006)
10. Babich, H., Sedletcaia, A., Kenigsberg, B.: In vitro cytotoxicity of protocatechuic acid to cultured human cells from oral tissue: involvement in oxidative stress. Pharmacol. Toxicol. **91**(5), 245–253 (2002)
11. Banks, C.E., Moore, R.R., Davies, T.J., Compton, R.G.: Investigation of modified basal plane pyrolytic graphite electrodes: definitive evidence for the electrocatalytic properties of the ends of carbon nanotubes. Chem. Commun. **16**, 1804–1805 (2004)
12. Banks, C.E., Crossley, A., Salter, C., Wilkins, S.J., Compton, R.G.: Carbon nanotubes contain metal impurities which are responsible for the "electrocatalysis" seen at some nanotube-modified electrodes. Angew. Chem. Int. Ed. **45**(16), 2533–2537 (2006)
13. Benitez, F.J., Beltran-Heredia, J., Acero, J.L., Gonzalez, T.: Degradation of protocatechuic acid by two advanced oxidation processes: ozone/UV radiation and H_2O_2UV radiation. Water Res. **30**(7), 1597–1604 (1996)
14. Borja, R., Banks, C., Maestro-Duran, R., Alba, J.: The effects of the most important phenolic constituents of olive mill wastewater on batch anaerobic methanogenesis. Environ. Technol. **17**(2), 167–174 (1996)
15. Brena, B., González-Pombo, P., Batista-Viera, F.: Immobilization of enzymes: a literature survey. In: Immobilization of Enzymes and Cells, pp. 15–31. Springer (2013)

16. Britto, P.J., Santhanam, K.S., Rubio, A., Alonso, J.A., Ajayan, P.M.: Improved charge transfer at carbon nanotube electrodes. Adv. Mater. **11**(2), 154–157 (1999)
17. Buchan, A., Collier, L.S., Neidle, E.L., Moran, M.A.: Key aromatic-ring-cleaving enzyme, protocatechuate 3,4-dioxygenase, in the ecologically important marineRoseobacter lineage. Appl. Environ. Microbiol. **66**(11), 4662–4672 (2000)
18. Camper, A.K., LeChevallier, M.W., Broadaway, S.C., McFETERS, G.A.: Growth and persistence of pathogens on granular activated carbon filters. Appl. Environ. Microbiol. **50** (6), 1378–1382 (1985)
19. Chan, W.-F., H-y, Chen, Surapathi, A., Taylor, M.G., Shao, X., Marand, E., Johnson, J.K.: Zwitterion functionalized carbon nanotube/polyamide nanocomposite membranes for water desalination. ACS Nano **7**(6), 5308–5319 (2013)
20. Corry, B.: Water and ion transport through functionalised carbon nanotubes: implications for desalination technology. Energy Environ. Sci. **4**(3), 751–759 (2011)
21. Das, R., Ali, M.E., Abd Hamid, S.B., Ramakrishna, S., Chowdhury, Z.Z.: Carbon nanotube membranes for water purification: a bright future in water desalination. Desalination **336**, 97–109 (2014). doi:10.1016/j.desal.2013.12.026
22. Datsyuk, V., Kalyva, M., Papagelis, K., Parthenios, J., Tasis, D., Siokou, A., Kallitsis, I., Galiotis, C.: Chemical oxidation of multiwalled carbon nanotubes. Carbon **46**(6), 833–840 (2008)
23. Di Paola, A., García-López, E., Marcì, G., Palmisano, L.: A survey of photocatalytic materials for environmental remediation. J. Hazard. Mater. **211**, 3–29 (2012)
24. Dumée, L.F., Sears, K., Schütz, J., Finn, N., Huynh, C., Hawkins, S., Duke, M., Gray, S.: Characterization and evaluation of carbon nanotube Bucky-Paper membranes for direct contact membrane distillation. J. Membr. Sci. **351**(1), 36–43 (2010)
25. Earth-Forum: Distribution of the world's water. Houston Museum of Natural Science. http://earth.rice.edu/mtpe/hydro/hydrosphere/hot/freshwater/0water_chart.html (2014). Accessed 26 Sept 2014
26. Eder, D.: Carbon nanotube—inorganic hybrids. Chem. Rev. **110**(3), 1348–1385 (2010)
27. Fang, H.-T., Liu, C.-G., Liu, C., Li, F., Liu, M., Cheng, H.-M.: Purification of single-wall carbon nanotubes by electrochemical oxidation. Chem. Mater. **16**(26), 5744–5750 (2004)
28. Feng, W., Ji, P.: Enzymes immobilized on carbon nanotubes. Biotechnol. Adv. **29**(6), 889–895 (2011)
29. Gadd, G.M., Griffiths, A.J.: Microorganisms and heavy metal toxicity. Microb. Ecol. **4**(4), 303–317 (1977)
30. Gao, Y., Kyratzis, I.: Covalent immobilization of proteins on carbon nanotubes using the cross-linker 1-ethyl-3-(3-dimethylaminopropyl) carbodiimide—a critical assessment. Bioconjug. Chem. **19**(10), 1945–1950 (2008)
31. Garcia, J., Gomes, H., Serp, P., Kalck, P., Figueiredo, J., Faria, J.: Carbon nanotube supported ruthenium catalysts for the treatment of high strength wastewater with aniline using wet air oxidation. Carbon **44**(12), 2384–2391 (2006)
32. Geng, Q., Guo, Q., Cao, C., Wang, L.: Investigation into NanoTiO_2/ACSPCR for decomposition of aqueous hydroquinone. Ind. Eng. Chem. Res. **47**(8), 2561–2568 (2008)
33. Gernjak, W., Krutzler, T., Glaser, A., Malato, S., Caceres, J., Bauer, R., Fernández-Alba, A.: Photo-Fenton treatment of water containing natural phenolic pollutants. Chemosphere **50**(1), 71–78 (2003)
34. Girelli, A.M., Mattei, E., Messina, A.: Phenols removal by immobilized tyrosinase reactor in on-line high performance liquid chromatography. Anal. Chim. Acta **580**(2), 271–277 (2006)
35. Gledhill, W.E.: Microbial toxicity and degradation test methodology: an industrial perspective. Tox. Assess. **2**(1), 89–96 (1987)
36. Goh, P., Ismail, A., Ng, B.: Carbon nanotubes for desalination: performance evaluation and current hurdles. Desalination **308**, 2–14 (2013)

37. Greń, I., Wojcieszyńska, D., Guzik, U., Perkosz, M., Hupert-Kocurek, K.: Enhanced biotransformation of mononitrophenols by *Stenotrophomonas maltophilia* KB_2 in the presence of aromatic compounds of plant origin. World J. Microbiol. Biotechnol. **26**(2), 289–295 (2010)
38. Gui, X., Wei, J., Wang, K., Cao, A., Zhu, H., Jia, Y., Shu, Q., Wu, D.: Carbon nanotube sponges. Adv. Mater. **22**(5), 617–621 (2010)
39. Guo, L., Morris, D.G., Liu, X., Vaslet, C., Hurt, R.H., Kane, A.B.: Iron bioavailability and redox activity in diverse carbon nanotube samples. Chem. Mater. **19**(14), 3472–3478 (2007)
40. Guzik, U., Hupert-Kocurek, K., Wojcieszyńska, D.: Intradiol dioxygenases—the key enzymes in xenobiotics degradation (2013)
41. Hemraj-Benny, T., Bandosz, T.J., Wong, S.S.: Effect of ozonolysis on the pore structure, surface chemistry, and bundling of single-walled carbon nanotubes. J. Colloid Interface Sci. **317**(2), 375–382 (2008)
42. Hinds, B.J., Chopra, N., Rantell, T., Andrews, R., Gavalas, V., Bachas, L.G.: Aligned multiwalled carbon nanotube membranes. Science **303**(5654), 62–65 (2004)
43. Hirose, Y., Tanaka, T., Kawamori, T., Olnishi, M., Makita, H., Mori, H., Satoh, K., Hara, A.: Chemoprevention of urinary bladder carcinogenesis by the natural phenolic compound protocatechuic acid in rats. Carcinogenesis **16**(10), 2337–2342 (1995)
44. Hossain, F., Perales-Perez, O.J., Hwang, S., Román, F.: Antimicrobial nanomaterials as water disinfectant: applications, limitations and future perspectives. Sci. Total Environ. **466**, 1047–1059 (2014)
45. Hou, P.-X., Liu, C., Cheng, H.-M.: Purification of carbon nanotubes. Carbon **46**(15), 2003–2025 (2008)
46. Hu, H., Yu, A., Kim, E., Zhao, B., Itkis, M.E., Bekyarova, E., Haddon, R.C.: Influence of the zeta potential on the dispersability and purification of single-walled carbon nanotubes. J. Phys. Chem. B **109**(23), 11520–11524 (2005)
47. Huang, W., Taylor, S., Fu, K., Lin, Y., Zhang, D., Hanks, T.W., Rao, A.M., Sun, Y.-P.: Attaching proteins to carbon nanotubes via diimide-activated amidation. Nano Lett. **2**(4), 311–314 (2002)
48. Hwang, E.T., Gu, M.B.: Enzyme stabilization by nano/microsized hybrid materials. Eng. Life Sci. **13**(1), 49–61 (2013)
49. IWMI: A Comprehensive Assessment of Water Management in Agriculture. In: Molden, D. (ed.) (2007)
50. Jia, G., Wang, H., Yan, L., Wang, X., Pei, R., Yan, T., Zhao, Y., Guo, X.: Cytotoxicity of carbon nanomaterials: single-wall nanotube, multi-wall nanotube, and fullerene. Environ. Sci. Technol. **39**(5), 1378–1383 (2005)
51. Jiang, K., Schadler, L.S., Siegel, R.W., Zhang, X., Zhang, H., Terrones, M.: Protein immobilization on carbon nanotubes via a two-step process of diimide-activated amidation. J. Mater. Chem. **14**(1), 37–39 (2004)
52. Kang, S., Pinault, M., Pfefferle, L.D., Elimelech, M.: Single-walled carbon nanotubes exhibit strong antimicrobial activity. Langmuir **23**(17), 8670–8673 (2007)
53. Kang, S., Herzberg, M., Rodrigues, D.F., Elimelech, M.: Antibacterial effects of carbon nanotubes: size does matter! Langmuir **24**(13), 6409–6413 (2008)
54. Kang, S., Mauter, M.S., Elimelech, M.: Physicochemical determinants of multiwalled carbon nanotube bacterial cytotoxicity. Environ. Sci. Technol. **42**(19), 7528–7534 (2008)
55. Kang, S., Mauter, M.S., Elimelech, M.: Microbial cytotoxicity of carbon-based nanomaterials: implications for river water and wastewater effluent. Environ. Sci. Technol. **43**(7), 2648–2653 (2009)
56. Kar, S., Bindal, R., Tewari, P.: Carbon nanotube membranes for desalination and water purification: challenges and opportunities. Nano Today **7**(5), 385–389 (2012)
57. Kar, S., Subramanian, M., Pal, A., Ghosh, A., Bindal, R., Prabhakar, S., Nuwad, J., Pillai, C., Chattopadhyay, S., Tewari, P.: Preparation, characterisation and performance evaluation of anti-biofouling property of carbon nanotube-polysulfone nanocomposite membranes. In:

CARBON MATERIALS 2012 (CCM12): Carbon Materials for Energy Harvesting, Environment, Nanoscience and Technology, vol. 1, pp. 181–185. AIP Publishing (2013)

58. Kaufmann, M., Melia-Teevan, K.: Turning the Tides of Crisis. http://blog.michellekaufmann.com/wp-content/uploads/2009/03/water_crisis.pdf (2009). Accessed July 2013
59. Kharraz, J.E., El-Sadek, A., Ghaffour, N., Mino, E.: Water scarcity and drought in WANA countries. Procedia Eng. **33**, 14–29 (2012)
60. Khin, M.M., Nair, A.S., Babu, V.J., Murugan, R., Ramakrishna, S.: A review on nanomaterials for environmental remediation. Energy Environ. Sci. **5**(8), 8075–8109 (2012)
61. Kim, K.-H., Ihm, S.-K.: Heterogeneous catalytic wet air oxidation of refractory organic pollutants in industrial wastewaters: a review. J. Hazard. Mater. **186**(1), 16–34 (2011)
62. Kim, J., Grate, J.W., Wang, P.: Nanostructures for enzyme stabilization. Chem. Eng. Sci. **61** (3), 1017–1026 (2006)
63. Klaine, S.J., Alvarez, P.J., Batley, G.E., Fernandes, T.F., Handy, R.D., Lyon, D.Y., Mahendra, S., McLaughlin, M.J., Lead, J.R.: Nanomaterials in the environment: behavior, fate, bioavailability, and effects. Environ. Toxicol. Chem. **27**(9), 1825–1851 (2008)
64. Kočí, K., Obalová, L., Lacný, Z.: Photocatalytic reduction of CO_2 over TiO_2 based catalysts. Chem. Pap. **62**(1), 1–9 (2008)
65. Kočí, K., Matějů, K., Obalová, L., Krejčíková, S., Lacný, Z., Plachá, D., Čapek, L., Hospodková, A., Šolcová, O.: Effect of silver doping on the TiO_2 for photocatalytic reduction of CO_2. Appl. Catal. B **96**(3), 239–244 (2010)
66. Kolaczkowski, S., Beltran, F., McLurgh, D., Rivas, F.: Wet air oxidation of phenol: factors that may influence global kinetics. Process Saf. Environ. Prot. **75**(4), 257–265 (1997)
67. Lam, C.-W., James, J.T., McCluskey, R., Hunter, R.L.: Pulmonary toxicity of single-wall carbon nanotubes in mice 7 and 90 days after intratracheal instillation. Toxicol. Sci. **77**(1), 126–134 (2004)
68. Lambert, J., Ajayan, P., Bernier, P., Planeix, J., Brotons, V., Coq, B., Castaing, J.: Improving conditions towards isolating single-shell carbon nanotubes. Chem. Phys. Lett. **226**(3), 364–371 (1994)
69. Lawrence, N.S., Deo, R.P., Wang, J.: Comparison of the electrochemical reactivity of electrodes modified with carbon nanotubes from different sources. Electroanalysis **17**(1), 65–72 (2005)
70. Li, J., Zhang, Y.: Cutting of multi walled carbon nanotubes. Appl. Surf. Sci. **252**(8), 2944–2948 (2006)
71. Li, P., Wang, X., Wang, H., Wu, Y.-N.: High performance liquid chromatographic determination of phenolic acids in fruits and vegetables. Biomed. Environ. Sci.: BES **6**(4), 389–398 (1993)
72. Li, Q., Mahendra, S., Lyon, D.Y., Brunet, L., Liga, M.V., Li, D., Alvarez, P.J.: Antimicrobial nanomaterials for water disinfection and microbial control: potential applications and implications. Water Res. **42**(18), 4591–4602 (2008)
73. Li, Y., Huang, X., Qu, Y.: A strategy for efficient immobilization of laccase and horseradish peroxidase on single-walled carbon nanotubes. J. Chem. Technol. Biotechnol. **88**(12), 2227–2232 (2013)
74. Ling, X., Wei, Y., Zou, L., Xu, S.: The effect of different order of purification treatments on the purity of multiwalled carbon nanotubes. Appl. Surf. Sci. **276**, 159–166 (2013)
75. Liu, J., Rinzler, A.G., Dai, H., Hafner, J.H., Bradley, R.K., Boul, P.J., Lu, A., Iverson, T., Shelimov, K., Huffman, C.B., Rodriguez-Macias, F., Shon, Y.S., Lee, T.R., Colbert, D.T., Smalley, R.E.: Fullerene pipes. Science **280**(5367), 1253–1256 (1998)
76. Liu, H., Ru, J., Qu, J., Dai, R., Wang, Z., Hu, C.: Removal of persistent organic pollutants from micro-polluted drinking water by triolein embedded absorbent. Bioresour. Technol. **100**(12), 2995–3002 (2009)
77. Liu, S., Wei, L., Hao, L., Fang, N., Chang, M.W., Xu, R., Yang, Y., Chen, Y.: Sharper and faster "nano darts" kill more bacteria: a study of antibacterial activity of individually dispersed pristine single-walled carbon nanotube. ACS Nano **3**(12), 3891–3902 (2009)

78. Liu, X., Wang, M., Zhang, S., Pan, B.: Application potential of carbon nanotubes in water treatment: a review. J. Environ. Sci. **25**(7), 1263–1280 (2013)
79. López, B.P., Merkoçi, A.: Improvement of the electrochemical detection of catechol by the use of a carbon nanotube based biosensor. Analyst **134**(1), 60–64 (2009)
80. Luo, J., Meyer, A.S., Mateiu, R.V., Pinelo, M.: Cascade catalysis in membranes with enzyme immobilization for multi-enzymatic conversion of CO_2 to methanol. New Biotechnol. **32**(3), 319–327 (2015)
81. Ma, P.-C., Kim, J.-K.: Carbon Nanotubes for Polymer Reinforcement. CRC Press (2011)
82. Majouga, A., Sokolsky-Papkov, M., Kuznetsov, A., Lebedev, D., Efremova, M., Beloglazkina, E., Rudakovskaya, P., Veselov, M., Zyk, N., Golovin, Y.: Enzyme-functionalized gold-coated magnetite nanoparticles as novel hybrid nanomaterials: synthesis, purification and control of enzyme function by low-frequency magnetic field. Colloids Surf., B **125**, 104–109 (2015)
83. Malawska, M., Ekonomiuk, A., Wiłkomirski, B.: Polycyclic aromatic hydrocarbons in peat cores from southern Poland: distribution in stratigraphic profiles as an indicator of PAH sources. Mires and Peat **1**(05), 1–14 (2006)
84. Masella, R., Cantafora, A., Modesti, D., Cardilli, A., Gennaro, L., Bocca, A., Coni, E.: Antioxidant activity of 3,4-DHPEA-EA and protocatecuic acid: a comparative assessment with other olive oil biophenols. Redox Rep. **4**(3), 113–121 (1999)
85. Matsushima, K., Kaneda, H., Harada, K., Matsuura, H., Hirata, K.: Immobilization of enzymatic extracts of *Portulaca oleracea* cv. roots for oxidizing aqueous bisphenol A. Biotechnol. Lett. **37**(5), 1037–1042 (2015)
86. Mauter, M.S., Elimelech, M.: Environmental applications of carbon-based nanomaterials. Environ. Sci. Technol. **42**(16), 5843–5859 (2008)
87. McCreery, R.L.: Advanced carbon electrode materials for molecular electrochemistry. Chem. Rev. **108**(7), 2646–2687 (2008). doi:10.1021/Cr068076m
88. Mestl, G., Maksimova, N.I., Keller, N., Roddatis, V.V., Schlögl, R.: Carbon nanofilaments in heterogeneous catalysis: an industrial application for new carbon materials? Angew. Chem. Int. Ed. **40**(11), 2066–2068 (2001)
89. Mubarak, N., Sahu, J., Abdullah, E., Jayakumar, N.: Removal of heavy metals from wastewater using carbon nanotubes. Sep. Purif. Rev. **43**(4), 311–338 (2014)
90. Mugdha, A., Usha, M.: Enzymatic treatment of wastewater containing dyestuffs using different delivery systems. Sci. Rev. Chem. Commun. **2**(1), 31–40 (2012). ISSN 2277 2669
91. Mukhopadhyay, A., Dasgupta, A.K., Chakrabarti, K.: Enhanced functionality and stabilization of a cold active laccase using nanotechnology based activation-immobilization. Bioresour. Technol. **179**, 573–584 (2015)
92. Nakamura, H., Nishikawa, A., Furukawa, F., Kasahara, K., Miyauchi, M., Son, H.Y., Hirose, M.: Inhibitory effects of protocatechuic acid on the post-initiation phase of hamster pancreatic carcinogenesis induced by N-nitrosobis (2-oxopropyl)amine. Anticancer Res. **20** (5B), 3423–3427 (2000)
93. Nakamura, Y., Torikai, K., Ohigashi, H.: Toxic dose of a simple phenolic antioxidant, protocatechuic acid, attenuates the glutathione level in ICR mouse liver and kidney. J. Agric. Food Chem. **49**(11), 5674–5678 (2001)
94. Narayan, R.J., Berry, C., Brigmon, R.: Structural and biological properties of carbon nanotube composite films. Mater. Sci. Eng., B **123**(2), 123–129 (2005)
95. Nepal, D., Geckeler, K.E.: pH-sensitive dispersion and debundling of single-walled carbon nanotubes: lysozyme as a tool. Small **2**(3), 406–412 (2006)
96. Ohnishi, M., Yoshimi, N., Kawamori, T., Ino, N., Hirose, Y., Tanaka, T., Yamahara, J., Miyata, H., Mori, H.: Inhibitory effects of dietary protocatechuic acid and costunolide on 7,12-dimethylbenz a anthracene-induced hamster cheek pouch carcinogenesis. Jpn. J. Cancer Res. **88**(2), 111–119 (1997)
97. Ören, A.H., Kaya, A.: Factors affecting adsorption characteristics of Zn^{2+} on two natural zeolites. J. Hazard. Mater. **131**(1), 59–65 (2006)

98. Osswald, S., Havel, M., Gogotsi, Y.: Monitoring oxidation of multiwalled carbon nanotubes by Raman spectroscopy. J. Raman Spectrosc. **38**(6), 728–736 (2007)
99. Paipetis, A., Kostopoulos, V.: Carbon Nanotube Enhanced Aerospace Composite Materials. Springer (2013)
100. Pang, R., Li, M., Zhang, C.: Degradation of phenolic compounds by laccase immobilized on carbon nanomaterials: diffusional limitation investigation. Talanta (2014)
101. Patel, Y., Gupte, A.: Biological treatment of textile dyes by agar-agar immobilized consortium in a packed bed reactor. Water Environ. Res. **87**(3), 242–251 (2015)
102. Peng, X., Sfeir, M.Y., Zhang, F., Misewich, J.A., Wong, S.S.: Covalent synthesis and optical characterization of double-walled carbon nanotube–nanocrystal heterostructures. J. Phys. Chem. C **114**(19), 8766–8773 (2010)
103. Pereira, M.G., Facchini, F.D.A., Filó, L.E.C., Polizeli, A.M., Vici, A.C., Jorge, J.A., Fernandez-Lorente, G., Pessela, B.C., Guisan, J.M., Polizeli, M.L.T.M.: Immobilized lipase from *Hypocrea pseudokoningii* on hydrophobic and ionic supports: determination of thermal and organic solvent stabilities for applications in the oleochemical industry. Process Biochem (2015)
104. Pulskamp, K., Diabaté, S., Krug, H.F.: Carbon nanotubes show no sign of acute toxicity but induce intracellular reactive oxygen species in dependence on contaminants. Toxicol. Lett. **168**(1), 58–74 (2007)
105. Pumera, M.: Carbon nanotubes contain residual metal catalyst nanoparticles even after washing with nitric acid at elevated temperature because these metal nanoparticles are sheathed by several graphene sheets. Langmuir **23**(11), 6453–6458 (2007)
106. Qu, X.L., Alvarez, P.J.J., Li, Q.L.: Applications of nanotechnology in water and wastewater treatment. Water Res. **47**(12), 3931–3946 (2013). doi:10.1016/j.watres.2012.09.058
107. Rao, G.P., Lu, C., Su, F.: Sorption of divalent metal ions from aqueous solution by carbon nanotubes: a review. Sep. Purif. Technol. **58**(1), 224–231 (2007)
108. Ren, X., Chen, C., Nagatsu, M., Wang, X.: Carbon nanotubes as adsorbents in environmental pollution management: a review. Chem. Eng. J. **170**(2), 395–410 (2011)
109. Riggs, J.E., Guo, Z., Carroll, D.L., Sun, Y.-P.: Strong luminescence of solubilized carbon nanotubes. J. Am. Chem. Soc. **122**(24), 5879–5880 (2000)
110. Ritter, L., Solomon, K., Forget, J., Stemeroff, M., O'leary, C.: A review of selected persistent organic pollutants. International Programme on Chemical Safety (IPCS) PCS/9539 Geneva: World Health Organization **65**, 66 (1995)
111. Rivas, F., Kolaczkowski, S., Beltran, F., McLurgh, D.: Development of a model for the wet air oxidation of phenol based on a free radical mechanism. Chem. Eng. Sci. **53**(14), 2575–2586 (1998)
112. Rivas, F.J., Frades, J., Alonso, M.A., Montoya, C., Monteagudo, J.M.: Fenton's oxidation of food processing wastewater components. Kinetic modeling of protocatechuic acid degradation. J. Agric. Food Chem. **53**(26), 10097–10104 (2005). doi:10.1021/jf0512712
113. Saleh, T.A.: The Role of Carbon Nanotubes in Enhancement of Photocatalysis (2013)
114. Saleh, T.A., Gupta, V.K.: Functionalization of tungsten oxide into MWCNT and its application for sunlight-induced degradation of rhodamine B. J. Colloid Interface Sci. **362** (2), 337–344 (2011)
115. Sarma, J., Mahiuddin, S.: Specific ion effect on the point of zero charge of α-alumina and on the adsorption of 3,4-dihydroxybenzoic acid onto α-alumina surface. Colloids Surf., A **457**, 419–424 (2014). doi:10.1016/j.colsurfa.2014.06.014
116. Savage, N., Diallo, M.S.: Nanomaterials and water purification: opportunities and challenges. J. Nanopart. Res. **7**(4–5), 331–342 (2005)
117. Shelimov, K.B., Esenaliev, R.O., Rinzler, A.G., Huffman, C.B., Smalley, R.E.: Purification of single-wall carbon nanotubes by ultrasonically assisted filtration. Chem. Phys. Lett. **282** (5), 429–434 (1998)
118. Shieh, Y.-T., Liu, G.-L., Wu, H.-H., Lee, C.-C.: Effects of polarity and pH on the solubility of acid-treated carbon nanotubes in different media. Carbon **45**(9), 1880–1890 (2007)

119. Shimp, R.J., Pfaender, F.K.: Effects of surface area and flow rate on marine bacterial growth in activated carbon columns. Appl. Environ. Microbiol. **44**(2), 471–477 (1982)
120. Smart, S., Cassady, A., Lu, G., Martin, D.: The biocompatibility of carbon nanotubes. Carbon **44**(6), 1034–1047 (2006)
121. Smith, B., Wepasnick, K., Schrote, K.E., Cho, H.-H., Ball, W.P., Fairbrother, D.H.: Influence of surface oxides on the colloidal stability of multi-walled carbon nanotubes: a structure–property relationship. Langmuir **25**(17), 9767–9776 (2009)
122. Subramanian, V., Wolf, E.E., Kamat, P.V.: Catalysis with TiO_2/gold nanocomposites. Effect of metal particle size on the Fermi level equilibration. J. Am. Chem. Soc. **126**(15), 4943–4950 (2004)
123. Subrizi, F., Crucianelli, M., Grossi, V., Passacantando, M., Pesci, L., Saladino, R.: Carbon nanotubes as activating tyrosinase supports for the selective synthesis of catechols. ACS Catal. **4**(3), 810–822 (2014). doi:10.1021/cs400856e
124. Suma, Y., Kim, D., Lee, J.W., Park, K.Y., Kim, H.S.: Degradation of catechol by immobilized hydroxyquinol 1,2-dioxygenase (1,2-HQD) onto single-walled carbon nanotubes. In: Proceedings of the International Conference on Chemical, Environmental Science and Engineering (ICEEBS'12) (2012)
125. Tanaka, T., Kojima, T., Kawamori, T., Yoshimi, N., Mori, H.: Chemoprevention of diethylnitrosamine-induced hepatocarcinogenesis by a simple phenolic acid protocatechuic acid in rats. Can. Res. **53**(12), 2775–2779 (1993)
126. Tanaka, T., Kojima, T., Suzui, M., Mori, H.: Chemoprevention of colon carcinogenesis by the natural product of a simple phenolic compound protocatechuic acid: suppressing effects on tumor development and biomarkers expression of colon tumorigenesis. Can. Res. **53**(17), 3908–3913 (1993)
127. Tanaka, T., Kojima, T., Kawamori, T., Mori, H.: Chemoprevention of digestive organs carcinogenesis by natural product protocatechuic acid. Cancer **75**(S6), 1433–1439 (1995)
128. Tseng, T.H., Wang, C.J., Kao, E.S., Chu, H.Y.: Hibiscus protocatechuic acid protects against oxidative damage induced by tert-butylhydroperoxide in rat primary hepatocytes. Chem. Biol. Interact. **101**(2), 137–148 (1996). doi:10.1016/0009-2797(96)03721-0
129. UN: Water scarcity. UN. http://www.un.org/waterforlifedecade/scarcity.shtml (2014). Accessed 4 Sept 2014
130. UNDP: One Planet to Share: Sustaining Human Progress in a Changing Climate. Asia-Pacific Human Development Report, New Delhi (2012)
131. Upadhyayula, V.K.K., Deng, S.G., Mitchell, M.C., Smith, G.B.: Application of carbon nanotube technology for removal of contaminants in drinking water: a review. Sci. Total Environ. **408**(1), 1–13 (2009). doi:10.1016/j.scitotenv.2009.09.027
132. U.S.-Census-Burea: International Data Base. http://web.archive.org/web/20060519101114/ http://www.geohive.com/global/geo.php?xml=hist3&xsl=hist3 (2014). Accessed 27 Sept 2014
133. Wang, H., Zhou, A., Peng, F., Yu, H., Yang, J.: Mechanism study on adsorption of acidified multiwalled carbon nanotubes to Pb (II). J. Colloid Interface Sci. **316**(2), 277–283 (2007)
134. Wepasnick, K.A., Smith, B.A., Schrote, K.E., Wilson, H.K., Diegelmann, S.R., Fairbrother, D.H.: Surface and structural characterization of multi-walled carbon nanotubes following different oxidative treatments. Carbon **49**(1), 24–36 (2011)
135. WWAP: The United Nations World Water Development Report 3: Water in a Changing World. UNESCO/Earthscan, Paris/London (2009)
136. WWAP: The United Nations World Water Development Report 4: Managing Water Under Uncertainty and Risk. UNESCO, Paris (2012)
137. Xu, R., Chi, C., Li, F., Zhang, B.: Laccase-polyacrylonitrile nanofibrous membrane: highly immobilized, stable, reusable, and efficacious for 2,4,6-trichlorophenol removal. ACS Appl. Mater. Interfaces **5**(23), 12554–12560 (2013)
138. Yan, H., Yang, X., Chen, J., Yin, C., Xiao, C., Chen, H.: Synergistic removal of aniline by carbon nanotubes and the enzymes of *Delftia* sp. XYJ6. J. Environ. Sci. **23**(7), 1165–1170 (2011)

139. Yang, S., Zhu, W., Li, X., Wang, J., Zhou, Y.: Multi-walled carbon nanotubes (MWNTs) as an efficient catalyst for catalytic wet air oxidation of phenol. Catal. Commun. **8**(12), 2059–2063 (2007)
140. Yang, J., Jiang, L.-C., Zhang, W.-D., Gunasekaran, S.: A highly sensitive non-enzymatic glucose sensor based on a simple two-step electrodeposition of cupric oxide (CuO) nanoparticles onto multi-walled carbon nanotube arrays. Talanta **82**(1), 25–33 (2010)
141. Yang, H.Y., Han, Z.J., Yu, S.F., Pey, K.L., Ostrikov, K., Karnik, R.: Carbon nanotube membranes with ultrahigh specific adsorption capacity for water desalination and purification. Nat. Commun. **4** (2013)
142. Yen, G.-C., Hsieh, C.-L.: Reactive oxygen species scavenging activity of Du-zhong (Eucommia ulmoides Oliv.) and its active compounds. J. Agric. Food Chem. **48**(8), 3431–3436 (2000)
143. Yu, J.-G., Zhao, X.-H., Yang, H., Chen, X.-H., Yang, Q., Yu, L.-Y., Jiang, J.-H., Chen, X.-Q.: Aqueous adsorption and removal of organic contaminants by carbon nanotubes. Sci. Total Environ. **482**, 241–251 (2014)
144. Yu, J.-G., Zhao, X.-H., Yu, L.-Y., Jiao, F.-P., Jiang, J.-H., Chen, X.-Q.: Removal, recovery and enrichment of metals from aqueous solutions using carbon nanotubes. J. Radioanal. Nucl. Chem. **299**(3), 1155–1163 (2014)
145. Zhang, L., Fang, M.: Nanomaterials in pollution trace detection and environmental improvement. Nano Today **5**(2), 128–142 (2010)
146. Zhang, M., Su, L., Mao, L.: Surfactant functionalization of carbon nanotubes (CNTs) for layer-by-layer assembling of CNT multi-layer films and fabrication of gold nanoparticle/CNT nanohybrid. Carbon **44**(2), 276–283 (2006)
147. Zhao, Z., Yang, Z., Hu, Y., Li, J., Fan, X.: Multiple functionalization of multi-walled carbon nanotubes with carboxyl and amino groups. Appl. Surf. Sci. **276**, 476–481 (2013)
148. Zhou, Q.-X., Wang, C.-Y., Fu, Z.-B., Tang, Y.-J., Zhang, H.: Effects of various defects on the electronic properties of single-walled carbon nanotubes: a first principle study. Front. Phys. **9**(2), 200–209 (2014)
149. Zhu, C., Luan, Z., Wang, Y., Shan, X.: Removal of cadmium from aqueous solutions by adsorption on granular red mud (GRM). Sep. Purif. Technol. **57**(1), 161–169 (2007)

Chapter 3
Carbon Nanotube Purification

Abstract Purification and functionalization of MWCNTs are challenging, but vital for their effective applications in various fields including CNTs based water purification technologies, catalysis, optoelectronics, biosensors, fuel cells, and electrode arrays. Existing CNT purification techniques often complicated and time-consuming, yielded shortened and curled MWCNTs that are not suitable for applications in certain fields such as membrane technologies, hybrid catalysis, optoelectronics, and sensor developments. Here we heeded the H_2O_2 synergetic actions with HCl and KOH in purifying and functionalizing pristine MWCNTs. The method (HCl/H_2O_2) annihilated all amorphous carbons and metal impurities from the pristine MWCNTs with a high purification yield (100%) compared with HCl alone (93.46%) and KOH/H_2O_2 (3.92%). We probed the findings using TEM, EDX, ATR-IR spectroscope, Raman spectroscope, and TGA analysis. The study is a new avenue for simple, rapid, low cost, and scalable purification of pristine MWCNTs for application in versatile fields.

3.1 Background

CNTs are one of the most fascinating nanomaterials having many attractive and useful physicochemical properties such as mechanical [26, 50], thermal [16], electrical [26, 42], and optoelectronic properties [3]. Since CNTs have defined by Iijima in 1991 [19], they have played significant shoot in many domains including novel water purification technologies [2, 7], polymer and composites [31, 40, 44], conductive cable fibers [28], hydrogen storage media [36], biomedical sciences [51], field emission displays [5], electrochemistry and nanosensors [24], nano-electrodes and microarrays [23], and many other versatile fields [5].

CNTs are often received or synthesized with other extraneous impurities such as non-nanotubes or amorphous carbons, ash, and metal catalysts with extreme hydrophobicity [17]. Existing CNT synthesis methods have added the impurities such as metal catalysts to increase the yield and reduce cost [22]. The quantity of these unwanted matters depends on the specific method used for CNT synthesis.

R. Das, *Nanohybrid Catalyst based on Carbon Nanotube*,
Carbon Nanostructures, DOI 10.1007/978-3-319-58151-4_3

Whatever might be the method of choice; impurities often hinder CNT performances, and confuse the understanding of their original functionalities and limiting their practical uses. For instance, CNT impurities have complicated and mislead observed electrocatalytic activities [4] and water pollutant's adsorption kinetics [1]. Further, CNT aggregation in various polymers and membranes [45] has destabilized its mechanical strengths and electrical conductivity [45]; making them unsuitable for water purifications.

HCl, H_2O_2, and KOH are easily available and simple wet chemicals in most of the ordinary laboratories. Here we treated MWCNTs with HCl [15, 37] as a reference. The synergistic effect of HCl/H_2O_2 mixture was systematically evaluated for improving the carbon yield and getting purified well graphitic layers of MWCNT. We also developed a new basic MWCNT purification technique involving a mixture of chemicals such as KOH/H_2O_2. Both treatments were effective for reacting with MWCNT impurities, but high purification yield might come from HCl/H_2O_2 rather than HCl alone and KOH/H_2O_2.

Both pristine and treated MWCNTs were characterized by various sensitive characterization tools. For instances, TEM was used to study the morphological changes of before and after treated MWCNTs. EDX was performed for the analyses of elemental composition and state of impurities. ATR-IR spectroscopy was conducted to estimate the degree and type of functionalization. Raman spectroscopy was employed to measure the defects and character of graphite bands. Finally, TGA was carried out to record the mass profiles of pristine and treated MWCNTs.

3.2 Literature Review

3.2.1 What are the Impurities Commonly Associated with the Pristine CNT?

CNTs as synthesized by AD, LA, and CVD methods must contained ash (most of them are carbonaceous), fullerenes, and metal catalysts (most of them are transition metals) as shown in Fig. 3.1. The amount of these impurities determine CNT

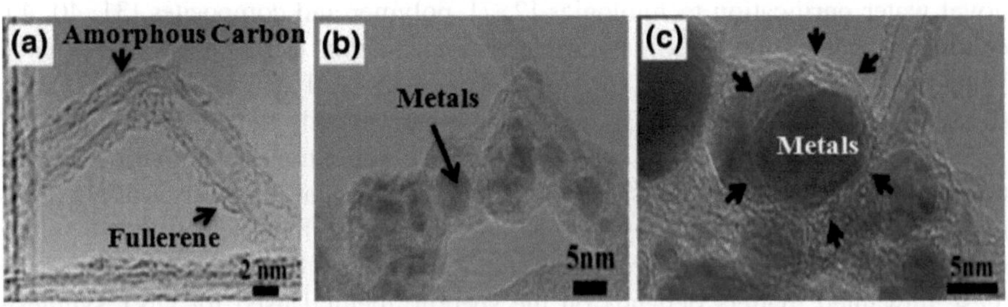

Fig. 3.1 TEM images of some pristine CNTs. Figures are adapted with permissions from Elsevier [34] and ACS [13]

diameter. Both the AD and LA add higher amount of impurities due to unvaporized graphitic particles, while CVD-mediated CNT have less impurities. Among these three impurity types, fullerene and amorphous carbon can be removed by organic washing and it is relatively easy. But removing transitional metals have remained a challenging task, since they are encapsulated by carbon layers. Therefore using acids and other agents need to break down the shells in order to get high purity CNT.

3.2.2 What Kinds of Characterization Methods are Commonly Used for CNT Impurities Detection?

In order to qualitatively and quantitatively measure the CNT impurities, several methods have mentioned in published studies, but it seems that there is no unique tool to determine the impurities in a precise and comprehensive way. Most of the microscopy units such as SEM, TEM, and AFM; TGA, Raman spectroscopy, and UV-vis-NIR spectroscopy are commonly used for this purpose. Electron microscopy (SEM, TEM, and AFM) can be used for visual detection of impurities and defects of CNT. But the microscopes are unable to quantify the amount of CNT impurities. In order to overcome this drawback, TGA has been used for quantitative detection of the quality of CNTs, especially the content of metal impurity. TGA should be conducted in air, because the oxidation temperature (>500 °C) always indicate pure carbon with less defects. Raman spectroscopy is a convenient technique for measuring the fraction of impurities using D/G ratio. But Raman spectroscope cannot give information on metal impurities. Finally, UV-vis-NIR spectroscopy is a commonly available unit in ordinary laboratory which can be used for rapid estimation of the relative purity of SWCNTs based on the integrated intensity of S22 transitions compared with that of a reference SWCNT sample. The drawback of this method is the difficulty in repeatedly preparing the standardized SWCNT film or solution and controlling film thickness or solution concentration, making it difficult for quantification analysis. Besides these above techniques, XPS is often used to determine the functional groups of CNTs. In addition, EDX is also used to semi-quantitatively identify the metal content in CNT samples.

3.2.3 What are the Methods Necessary for CNT Purification?

Three types of CNT purification methods are commonly used: (a) physical, (b) chemical, and (c) combination of physical and chemical methods. The physical methods are filtration, centrifugation, high-temperature annealing, etc. The methods remove the impurities from the CNTs based on the differences in their magnetism, gravity in solution, aspect ratio, and physical size. Physical methods commonly used for impure graphitic sheets removal from CNT. Since the methods do not involve any chemical agents, the original CNT integrity shall be reserved. But most of the physical methods are less effective due to time-consuming and complicated procedures. Second, the chemical methods oxidize the CNT for impurities removal. Many oxidation techniques such as gas phase oxidation, liquid phase oxidation, and electrochemical oxidation are commonly used chemical methods for CNT oxidation. The methods oxidize carbonaceous impurities and dissolute the metal catalyst on CNTs in presence of acids. Chemical methods are effective than physical methods, but it influences the structure of CNTs: create defected CNTs. Therefore, combined treatments of physical and chemical are often required. Such hybrid approach can increase highly purified CNT yield and high-quality CNT products. The methods should be used in such a way that the basic morphology and structure of CNTs should be preserved with desired purity. Figure 3.2 reveals a flowchart by using that one can obtain the desired CNT purity by using representative procedures.

3.2.4 What Challenges Does Exist with CNT Purification?

Although a substantial progress has been made for developing CNT purification methods, there is no unique method for yielding high-quality CNT of low cost and short time involved. Most of the methods are not reproducible, eco-friendly, and scaling up which should be overcome. Other challenges are given as bullet points below:

- Large-scale purification of CNT with desired diameter, chirality, and thickness is absent.
- Conductivity of CNT should be preserved, since most of the chemical methods oxidize the CNT skeleton.
- Specific method is required to develop for fabricating specific purified CNT-based technology.
- Scaling phenomena such as uniformity of CNT and its homogeneous contact between impurities and oxidants should be taken into account.
- Establishing novel characterization tools is required to evaluate and compare different CNT samples.
- Establishing a standard protocol that can completely describe the features of the purified CNTs, such as CNT content, amounts of different impurities, defects, etc.

In order to tackle these above challenges, here we developed some novel chemical methods for CNT purification. HCl has been used to purify CNTs [15, 37], but often incapable to the complete removal of metal impurities [48]. On the

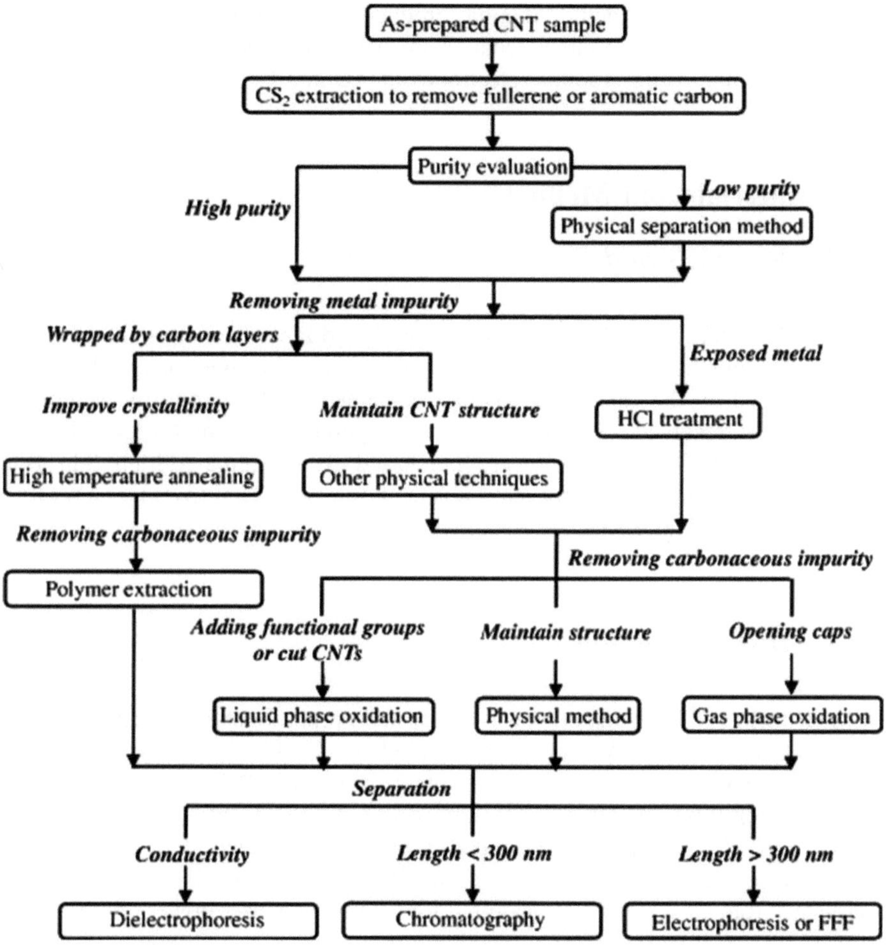

Fig. 3.2 A flowchart showing the representative procedures based on the function and characteristics of the purifications. Adapted with permission from Elsevier [17]

other hand, pristine few walled carbon nanotubes (FWCNTs) have been oxidized and purified by H_2O_2 [14], but the method has thus far not extensively studied for MWCNT purification except a few [32, 49]. Although SWCNTs have been purified by a mixture of HCl and H_2O_2 [20, 21, 41, 48, 52], no study has yet been adopted for MWCNT purification by using the HCl/H_2O_2 mixture. As a corollary, here we selected the method for purifying MWCNTs and observed high purification yield. Basic treatment involving NH_4OH/H_2O_2 has been found capable in purifying MWCNTs [8]. Consequently, we developed another novel mixture of KOH/H_2O_2, but it showed fewer competencies to dissolute metals from the pristine MWCNTs. We focused on MWCNTs because of their low preparation cost and availability in large quantities. This might be because of the development of low-cost simple technology called CVD method that has produced CNTs using cheap catalysts (Fe, Ni, Co, etc.), and carbon sources (coal, charcoal, asphalt, etc.) and provided higher yields rather than using graphitic electrodes in AD and LA methods [33].

3.3 Materials and Methods

3.3.1 *Materials and Reagents*

Pristine MWCNTs of 13 ± 5 and 4 nm in outer and inner diameters and >1 μm in length were purchased from Bayer Material Science AG (Germany). The tubes were prepared by the catalytic CVD and contained >95% carbon by weight and were used as received. HCl and H_2O_2 were purchased from Merck Sdn Bhd. (Malaysia). KOH and ethanol were purchased from Sigma-Aldrich Sdn Bhd. (Malaysia). The purity of all reagents was ≥ 99% except HCl (37%); H_2O_2 (30%) and ethanol (70%) in water.

3.3.2 *Instrumentation*

TEM (Hitachi-HT7700, Japan) was used for the morphological characterizations of the MWCNTs at 120 kV. An EDX coupled with a FE-SEM (QUANTA FEG 450, FEI, USA) was used for tracing the elements of MWCNTs. An X-Max Silicon Drift detector (Oxford, UK) of 80 mm^2 was used to identify the elements and energy, and relative intensity of emitted X-rays was analyzed at 10 keV. ATR-IR spectra were recorded on a KBr using an IR spectrometer (IFS 66 v/S, Bruker, Germany). Raman spectra were acquired for 10 min at a laser power of 100 on Ar^+ laser (514 nm) focused (50X objective) to a spot size of about 1.5–2.0 μm (Renishaw inVia, UK). TGA (TGA/SDTA 851, Mettler Toledo, USA) was performed under air flow (50 ml) between 25 and 1000 °C at 10 °C/min.

3.3.3 *Wet Chemical Treatments of MWCNT*

Three wet chemical treatments were performed to purify and oxidize the as-obtained pristine MWCNTs:

(a) **HCl Treatment**: Pristine MWCNT (0.5 g) was treated with 100 ml of HCl (36 wt%) and sonicated at 50 °C for 5 h in an ultrasonication bath (Series 400; Powersonic, 40 kHz; Korea) [37]. The method was as a reference.

(b) **HCl/H_2O_2 Treatment**: Pristine MWCNT (0.5 g) was dispersed into a 25 ml mixture (70:30) of HCl (36 wt%) and H_2O_2 (30 wt%), and the final mixture was sonicated for 5 h at 50 °C, 40 kHz.

(c) **KOH/H_2O_2 Treatment**: Pristine MWCNT (0.5 g) was dispersed into a 20 ml mixture (50:50) of KOH (25 wt%) and H_2O_2 (30 wt%), and the mixture was sonicated for 5 h at 50 °C, 40 kHz.

3.3.3.1 Removal of Residual Impurities

All treated MWCNTs were extracted from the residual acids, and bases, metallic by-products, and carbonaceous impurities by repeated cycle of dilutions and centrifugations [dissolved in 1.0 L of deionized water and centrifuged (Eppendorf-5430R; Germany) at 7000 rpm for 30 min]. The supernatant was carefully collected when the MWCNTs were precipitated at the bottom of the polyethylene centrifuge tube. The procedure was repeated 5–6 times until the resistivity of the supernatant was greater than 0.5 MΩ cm and pH was ~7.0. The treated MWCNTs were then rinsed with ethanol (70 wt%) and dried overnight in a vacuum oven at 100 °C.

3.3.3.2 Stock Solution Preparation and Solubility Measurement

About 1.0 mg of pristine and treated MWCNTs was added separately into 5.0 ml of water (pH 7.1) in order to measure their colloidal stability and solubility. The solutions were then sonicated for 30 min at 40 kHz to get stable colloidal suspensions. Within 30 min all MWCNTs were dispersed in such a way that no visible aggregates were observed in water vials. Finally, undesirable dusts were removed carefully, and every solution was left for an hour to see the dispersibility and colloidal stability of MWCNTs.

3.4 Results

3.4.1 *Predicted Chemical Reactions of HCl, H_2O_2, and KOH with MWCNT*

In order to get pure MWCNTs, methods such as HCl, HCl/H_2O_2, and KOH/H_2O_2 were found promising. Herein we proposed the mechanisms of purifying MWCNTs using HCl, HCl/H_2O_2, and KOH/H_2O_2 as shown in Fig. 3.3. The method (HCl/H_2O_2) can purify MWCNTs through different routes. The metals that are usually present in pristine CNTs could act as catalysts to produce hydroxyl radical

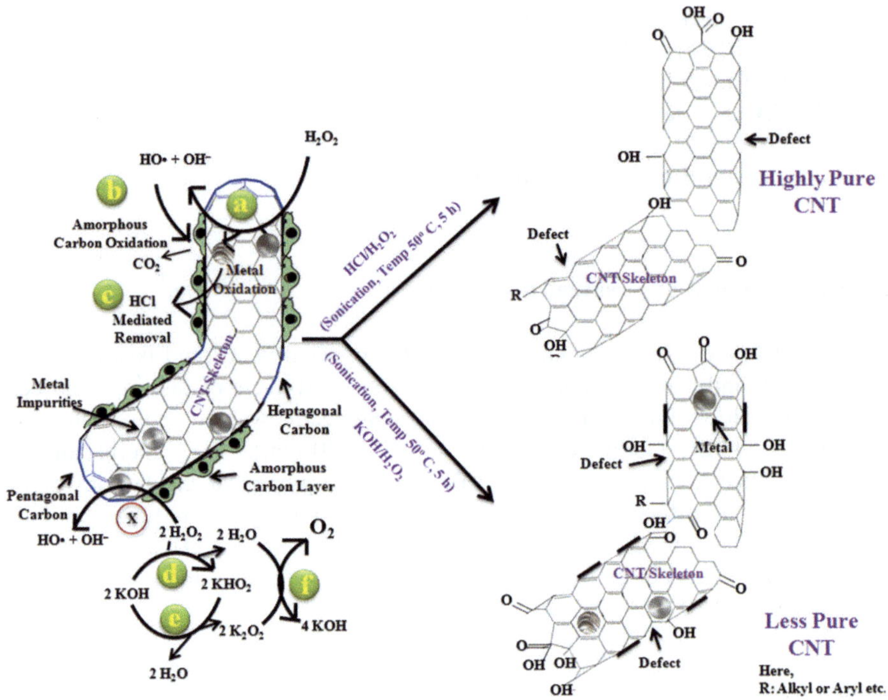

Fig. 3.3 Schematic representation of the localized catalytic reactions of (**a–c**) HCl/H_2O_2 and (**d–f**) KOH/H_2O_2 with pristine MWCNT

($OH^{\cdot}$) through Fenton's chemistry [47] (Fig. 3.3a), which is a stronger oxidizing agent than H_2O_2 as shown in Eq. 3.1.

$$H_2O_2 + \text{metal (reduced)} \rightarrow OH^{\cdot} + OH^{-} + \text{metal (oxidized)} \tag{3.1}$$

The radical ($OH^{\cdot}$) is then reacted with amorphous carbon impurities of pristine CNTs [49] and converted them into CO_2 [48] (Fig. 3.3b) as stated in Eq. 3.2.

$$4\,OH^{\cdot} + C \rightarrow CO_2 + 2H_2O \tag{3.2}$$

The oxidized metals and other impurities are then dissolved into HCl (Fig. 3.3c) and subsequently removed them through repetitive filtering and washing solutions. However, a mixture of KOH and H_2O_2 was unable to complete amorphous carbon oxidation and removal of metal impurities from MWCNT. This is because of the chemical reactions between KOH and H_2O_2 (Fig. 3.3d–f). Herein KOH decreases the availability of H_2O_2 in the system, so there is least chance to produce free radicals and other etching agents. The ultimate products of the reaction might be the

KOH and O_2. Oxygen has evaporated while KOH may have some chemical interactions with amorphous carbons which might be negligible to remove MWCNT core impurities.

3.4.2 TEM Analysis

TEM microscope was used to closely examine the contents of amorphous carbon and trace metal catalysts trapped into the tubular interstitial spaces in pristine and treated MWCNTs as shown in Fig. 3.4. Pristine MWCNTs reflected clumped, cloudy, and blackish amorphous carbon-containing MWCNTs (Fig. 3.4a). The impure carbonaceous particles were found to be wrapped around the nanotube structures, and metal catalysts were trapped into the MWCNTs. Most of the HCl-treated MWCNTs were thick in sizes (Fig. 3.4b). The HCl/H_2O_2 treatment produced much cleaner, thin, fresh, and complete amorphous carbon and metals-free MWCNTs, and the nanotubes were appeared in thin and loosely connected bundles (Fig. 3.4c). In the KOH/H_2O_2 treatment, some of the amorphous carbons from MWCNT surfaces were removed. However, the effect was local since the presence of some non-nanotube carbon impurities was globally obvious (Fig. 3.4d). The nanotubes were appeared as more flattened and thick than those of HCl (Fig. 3.4b) and HCl/H_2O_2 (Fig. 3.4c) treated MWCNTs.

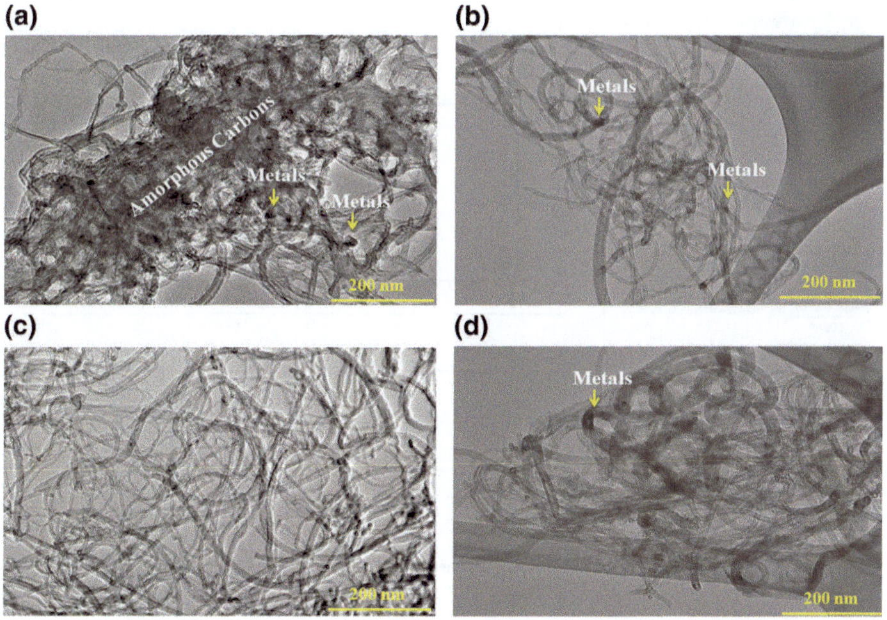

Fig. 3.4 TEM images of **a** pristine; **b** HCl; **c** HCl/H_2O_2 and **d** KOH/H_2O_2 treated MWCNTs

3.4.3 EDX Analysis

EDX is an important characterization tool for measuring the extent of CNT oxidation and elemental composition [49]. EDX findings of the pristine and treated MWCNTs are shown in Fig. 3.5 and representative analysis is listed in Table 3.1.

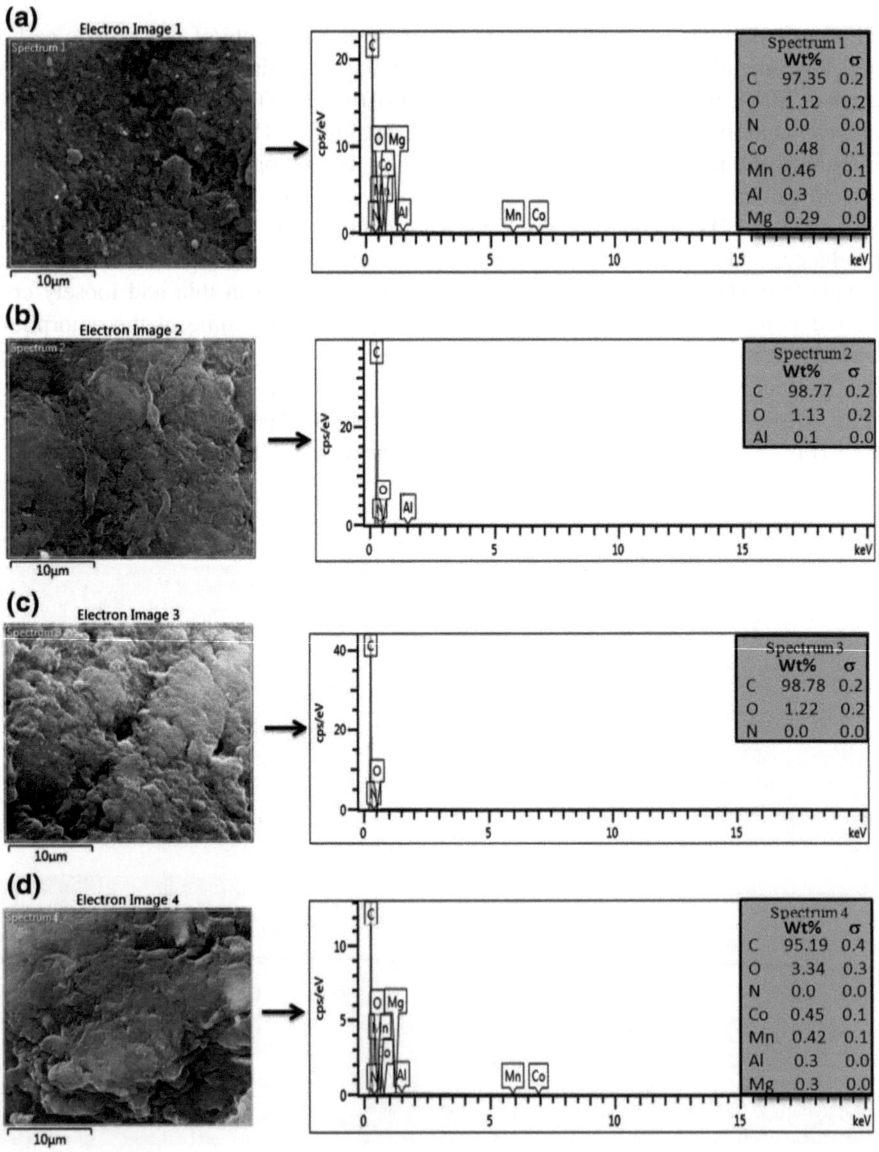

Fig. 3.5 EDX profiles of **a** pristine, **b** HCl, **c** HCl/H_2O_2 and **d** KOH/H_2O_2 treated MWCNTs

Table 3.1 Elemental analysis before and after MWCNT treatments in this study

Specimen	Treatment time (h)	Elemental composition (wt%)							
		C	O	Co	Mn	Al	Mg	Sum	Purification yield (%)
Pristine-MWCNT	0	97.35	1.12	0.48	0.46	0.3	0.29	100	–
MWCNT-HCl	5	98.77	1.13	0	0	0.1	0	100	93.46
MWCNT-HCl/H_2O_2	5	98.78	1.22	0	0	0	0	100	100
MWCNT-KOH/H_2O_2	5	95.19	3.34	0.45	0.42	0.3	0.3	100	3.92

Purification yields of the pristine and treated MWCNTs were calculated based on following Eq. 3.3.

$$\text{Purification yield (\%)} : W_0 - W_t / W_0 \times 100 \tag{3.3}$$

where W_0 is the metal content of pristine MWCNT (%), and W_t is the metal content of purified MWCNT (%).

According to Table 3.1 and Fig. 3.5; by far, the largest element in the as-received pristine MWCNT was carbon (either graphitic or amorphous), with some extent of oxygen (Fig. 3.5a). However, pristine MWCNTs were highly contaminated with metal impurities such as Co, Mn, Al and Mg (Fig. 3.5a). After wet chemical agent treatments, it was found that the quantity of graphitic carbons was slightly increased with good purification yields of 93.46 (HCl) and 100% (HCL/H_2O_2) (Fig. 3.5b, c). In contrast, KOH/H_2O_2 was incapable to complete removal of metal impurities and showed lowest purification yield of 3.92% (Fig. 3.5d).

3.4.4 ATR-IR Analysis

ATR-IR spectroscopy was performed for characterizing the functionalities produced following wet chemical treatments (HCl, HCl/H_2O_2 and KOH/H_2O_2) resulting MWCNT purifications. The IR spectra of the pristine and treated MWCNTs are shown in Fig. 3.6. The dominant IR spectrum at 3409 cm^{-1} was assigned to the stretching vibrations of intermolecularly hydrogen bonded OH:OH groups (Fig. 3.6a–d) [29, 30].

The IR transmittance peak at 2907 cm^{-1} which was dominant in HCl/H_2O_2 (Fig. 3.6c) and KOH/H_2O_2 (Fig. 3.6d) treated MWCNTs was assigned to sp^2 and/or sp^3 C–H stretching vibrations [39]. The transmittance bands at 2422 and 2279 cm^{-1} were observed for pristine (Fig. 3.6a), HCl/H_2O_2 (Fig. 3.6c) and KOH/H_2O_2 (Fig. 3.6d), but absent in HCl treated MWCNTs (Fig. 3.6b), pointed out the grafting of some –CO functionalities [27]. The peak at 1630 cm^{-1} (Fig. 3.6a–d) was due to the stretching vibration of either C=C [6] or C=O of quinone [27]. The prominent peak at 1044 cm^{-1} (Fig. 3.6a–d) was due to –OH

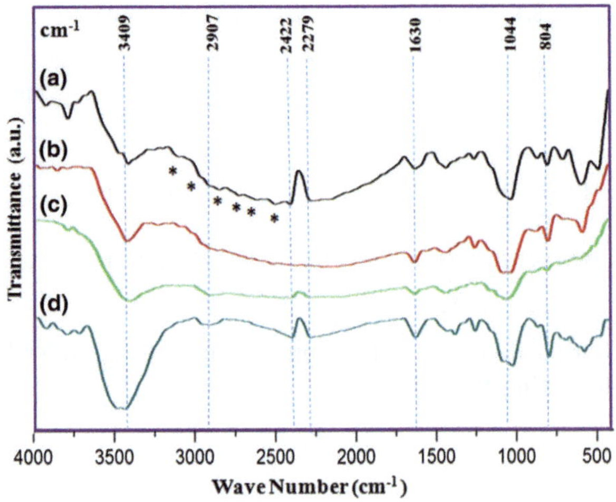

Fig. 3.6 ATR-IR spectra of **a** pristine, **b** HCl, **c** HCl/H_2O_2 and **d** KOH/H_2O_2 treated MWCNTs

group generated either by the atmospheric oxidation or oxidized by wet chemical treatments [27]. In addition, a peak that appeared at 804 cm^{-1} (Fig. 3.6a–d) was due to epoxy and oxirane rings evolved from C–O– groups and ring deformation of substituted aromatic structures [6].

3.4.5 Raman Spectroscopy Analysis

Typically MWCNTs represent two significant high frequency bands called D and G- bands at 1330 and 1585 cm^{-1} for CNT structural defects and graphite in-plane vibration, respectively [9, 11]. In this study, D and G- bands appeared at ~1349 and ~1588 cm^{-1} both in pristine and treated MWCNTs, respectively (Fig. 3.7). It shows the D-band intensities were practically constant in both pristine and treated MWCNTs. In contrast, G-band intensities were significantly increased in treated MWCNTs especially for HCl/H_2O_2 treated MWCNTs.

Finally, the purity states of the pristine and treated MWCNTs were compared from the intensity ratio of the G (I_G) and D bands (I_D) [8]. High I_G/I_D ratio meant well graphitic CNTs, whereas low I_G/I_D indicated contaminated CNTs [8, 14]. The highest I_G/I_D ratio (0.91) was found for HCl/H_2O_2 treated MWCNTs, whereas the ratios were for HCl (0.81) and KOH/H_2O_2 (0.73) treated MWCNTs (Fig. 3.7).

3.4.6 TGA Analysis

TGA was performed to measure the amorphous carbons; oxidation defects and overall quality of purified MWCNTs. TGA of pristine and treated MWCNTs with

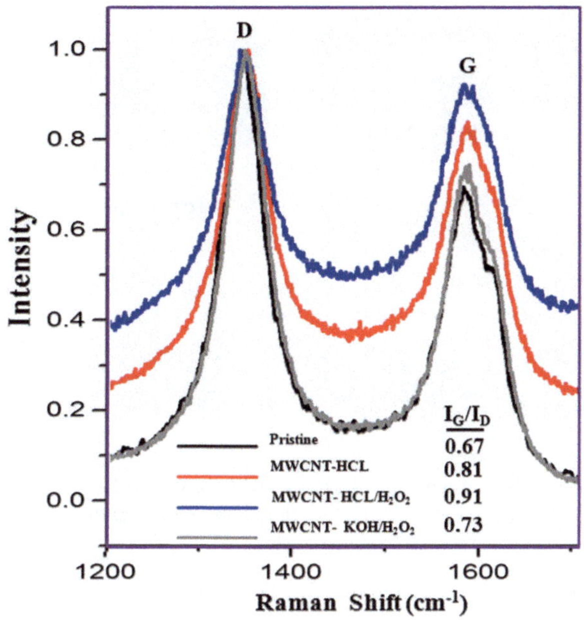

Fig. 3.7 Normalized Raman spectra of pristine and treated MWCNTs

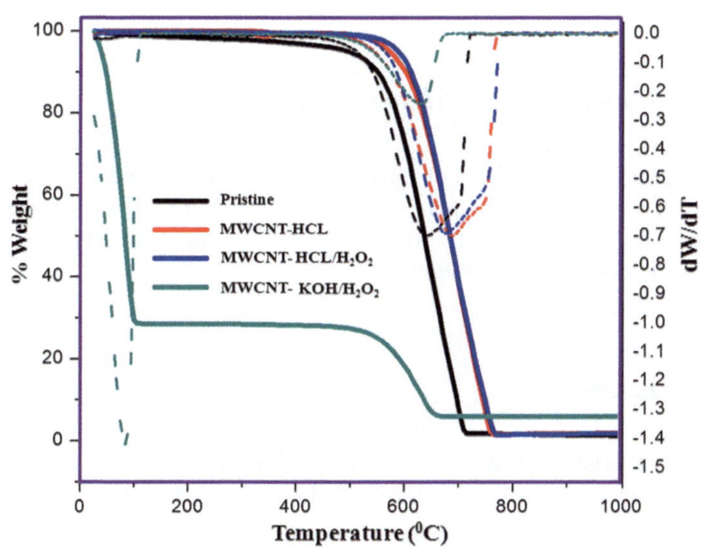

Fig. 3.8 TGA (*solid lines*) and DTA (*dashed lines*) of pristine and treated MWCNTs

their differential thermogravimetric analysis (DTG) spectra are presented in Fig. 3.8. At first, the pristine and KOH/H_2O_2 treated MWCNTs showed lowest decomposition temperature at around 100 °C, attributing to the pyrolytic evolution

Fig. 3.9 Solubility profiles of **a** pristine, **b** HCl, **c** HCl/H_2O_2 and **d** KOH/H_2O_2 treated MWCNTs. All MWCNTs were dissolved in water at concentrations of 0.2 mg/ml

of hydroxyl and/or water molecules [8]. TGA of pristine and KOH/H_2O_2-treated MWCNTs showed highest decomposition temperatures at 650 and 700 °C, respectively (Fig. 3.8). In contrasts, TGA of HCl and HCl/H_2O_2-treated MWCNTs started to oxidize at around 600 °C and reached final oxidation at 750 °C (Fig. 3.8).

3.4.7 Colloidal Stability Measurement

Figure 3.9 represents the camera observations of the colloidal dispersibility of MWCNTs in water solvents of pH~7.0. It shows less solubility of pristine, HCl and KOH/H_2O_2 treated MWCNTs (Fig. 3.9a–d). In contrast, HCl/H_2O_2-treated MWCNTs showed moderate solubility.

3.5 Discussion

Purification of pristine MWCNTs using HCl, HCl/H_2O_2, and KOH/H_2O_2 was found promising with certain variations. In Fig. 3.4, TEM image (b) shows a few HCl-treated MWCNTs were thick in sizes which suggest small percentages of non-nanotube carbonaceous agents and/or metal impurities might be present in MWCNTs. In contrast, HCl/H_2O_2 mixture produced thin thread like MWCNT structures and simultaneously maintaining intact nanotube integrity (Fig. 3.4c). Although the oxidizing strength of H_2O_2 is high (pKa 11.6), it did not produce vigorous CNT fragmentations upon the purification process (Fig. 3.4c). So the method could overcome the drawbacks of existing CNT purification methods such as HNO_3, H_2SO_4, and so on [8, 17, 25]. HNO_3-treated CNTs have shown to be degraded and chopped into smaller pieces [17]. Such shortened CNT fragments would have limited applications and are unsuitable for water purification technologies. On the other hand, KOH/H_2O_2 mixture was found to be ineffective to clean MWCNTs, since most of the impurities remained attached globally that made thick and slightly blackish MWCNT surfaces (Fig. 3.4d).

EDX analysis confirmed that the atomic carbon which should be 100% in pure MWCNTs without defects was 1.53% less than the calculated value, indicating equal amounts of metal impurities present in pristine MWCNTs (Fig. 3.5a and Table 3.1). The presence of 1.12% oxygen suggested some unintentional oxidations of pristine MWCNTs probably by environmental ozone and hydroxyl radicals [12, 46]. Treating MWCNTs with HCl and HCl/H_2O_2 was found promising for high yield purifications because of increasing overall carbon, and oxygen percentages and eliminating metal impurities from the contaminated CNTs (Fig. 3.5b, c). But the efficiency of HCl/H_2O_2 was higher than HCl alone in removing all metal ions. This is because of synergistic actions of H_2O_2 to produce $OH^{\bullet}$ that might help to dissolute metals from pristine MWCNTs completely (Fig. 3.3a–c). Compared with HCl (Fig. 3.5b) and HCl/H_2O_2 treated MWCNTs (Fig. 3.5c), KOH/H_2O_2 treatment was least effective one to cut out metal contaminants (Fig. 3.5d) because of its inability to produce any oxidizing agents or free radicals (Fig. 3.3d–f). The results are in consistent with TEM analyses described in Sect. 3.4.2. Therefore, we hypothesized that the HCl/H_2O_2 mixture can be an effective method for the complete purification of pristine MWCNTs with intact nanotube skeleton than the HCl alone and KOH/H_2O_2 treatments.

Compared with pristine (Fig. 3.6a), HCl (Fig. 3.6b), and HCl/H_2O_2-treated MWCNTs (Fig. 3.6c), a major, high-intense, and broadening of IR spectrum in KOH/H_2O_2-treated MWCNTs (Fig. 3.6d) at 3409 cm^{-1}, attributing to form more –OH groups [6] as shown in EDX analysis (Fig. 3.5d). The group might be contributed to the high oxygen percentage of KOH/H_2O_2 treated MWCNTs. Some weak peaks that appeared at 2500–3500 cm^{-1} region in pristine MWCNTs (shown by asterisks in Fig. 3.6a) were disappeared following chemical treatments (Fig. 3.6b–d). It revealed the presence of some minor functional groups of the pristine MWCNTs anchored by amorphous carbons and other non-nanotube carbon fragments, which were successfully removed upon wet purification agent treatments.

The intensity of the D-band, which is induced by nonzero center phonon mode usually depends on the presence of disordered carbon atomic networks in Raman spectroscopy as shown in Fig. 3.7 [10, 21]. Constant D-band intensities of all treated MWCNTs indicated less defected purified MWCNTs (Fig. 3.7). This might be due to the milder reaction conditions among graphitic carbons of MWCNTs with HCl, HCl/H_2O_2, and KOH/H_2O_2. The etching properties of $OH^{\bullet}$, which was generated by Fenton's chemistry [47] may have direct affinity to oxidize amorphous carbons due to the presence of many active sites on it [8] (Fig. 3.3b) rather than oxidizing graphitic layer's carbon atoms. On the other hand, KOH/H_2O_2 was unable to directly react with graphitic skeleton, since most of the amorphous carbons were wrapped around the pristine MWCNTs (Fig. 3.3d–f). However, increased G-band intensity and high I_G/I_D ratio of HCl/H_2O_2 treated MWCNTs suggesting the better efficiency of the mixture in removing amorphous and carbonaceous materials from MWCNTs [14] in contrasts to HCl and KOH/H_2O_2 methods [10]. As a result, we hypothesized that HCl/H_2O_2 not only be able to purify MWCNTs completely, but it can maintain intact MWCNT integrity which is necessary for the development of effective catalysts, sensor and optoelectronic devices.

TGA is an effective characterization method to quantitatively evaluate the content of CNTs [17] as shown in Fig. 3.8. Herein an oxidation temperature means the temperature where MWCNTs lose its weight and thus show the highest derivative weight curve. It reflects the stability of MWCNTs at a given temperature. Typically, amorphous carbons oxidized at lower temperature regions such as around 500 °C [18] due to their lower activation energy and the presence of many heat sensitive active sites [8]. In contrast, pure graphitic carbons might decompose at higher temperature ranges [35]. The fastest decomposition temperatures of HCl and HCl/H_2O_2 treated MWCNTs were moved to the highest temperatures as compared with pristine and KOH/H_2O_2 treated MWCNTs (Fig. 3.8). It suggests the efficiency of HCl and HCl/H_2O_2 than KOH/H_2O_2 treated MWCNTs in purifying contaminated MWCNTs [35].

Although the purification yields of MWCNTs were considerably improved by treating with HCl, HCl/H_2O_2 and KOH/H_2O_2 treatments, they might have less efficiency to produce dispersed MWCNTs in water solvents (Fig. 3.9). The pristine MWCNTs have a strong tendency to precipitate in polar solvents (Fig. 3.9a). This is because of their hydrophobic graphite sheets and strong surface energies [43]. Functionality such as –COOH has found effective to make electrostatic interactions needed to form a stable CNT colloidal suspension [38]. Insolubility of treated MWCNTs indicated that the group was not sufficiently anchored on MWCNT surfaces by HCl, HCl/H_2O_2 and KOH/H_2O_2 treatments. HCl/H_2O_2 treated MWCNTs were slightly dispersed might be because of some nucleophilic reactions of C=O on MWCNT surfaces and water molecules. Although KOH/H_2O_2 treated MWCNTs showed higher oxygen percentage as shown in Fig. 3.5d, most of the groups were might in the form of hydroxyl, oxirane and epoxy rather than –COOH groups, making them insoluble in water.

3.6 Conclusions

Purification and functionalization of MWCNTs are challenging, but vital for their effective applications in various fields including CNTs-based water purification technologies, catalysis, optoelectronics, biosensors, fuel cells, and electrode arrays. Existing CNT purification techniques are often complicated and time-consuming, yielded shortened and curled MWCNTs that are not suitable for applications in certain fields such as membrane technologies, hybrid catalysis, optoelectronics, and sensor developments. Here we heeded the H_2O_2 synergetic actions with HCl and

KOH in purifying and functionalizing the pristine MWCNTs. The method (HCl/H_2O_2) annihilated all amorphous carbons and metal impurities from the pristine MWCNTs and produced well-ordered and thread-like MWCNT graphitic skeleton with a high purification yield (100%) compared with HCl alone (93.46%) and KOH/H_2O_2 (3.92%). The methods HCl, HCl/H_2O_2, and KOH/H_2O_2 were not corrosive enough in MWCNT defecting and also were incapable to anchor appropriate –COOH functionality to solubilize them in water. Thus, the MWCNT purities depend on the nature of wet chemical agents and impurities present in the MWCNT structure.

References and Future Readings

1. Agnihotri, S., Mota, J.P., Rostam-Abadi, M., Rood, M.J.: Adsorption site analysis of impurity embedded single-walled carbon nanotube bundles. Carbon **44**(12), 2376–2383 (2006)
2. Ali, M., Das, R., Maamor, A., Hamid, S.B.A.: Multifunctional Carbon Nanotubes (CNTs): a new dimension in environmental remediation. Adv. Mater. Res. **832**, 328–332 (2014)
3. Avouris, P., Chen, J.: Nanotube electronics and optoelectronics. Mater Today **9**(10), 46–54 (2006)
4. Banks, C.E., Crossley, A., Salter, C., Wilkins, S.J., Compton, R.G.: Carbon nanotubes contain metal impurities which are responsible for the "Electrocatalysis" Seen at some nanotube-modified electrodes. Angew. Chem. Int. Ed. **45**(16), 2533–2537 (2006)
5. Baughman, R.H., Zakhidov, A.A., de Heer, W.A.: Carbon nanotubes–the route toward applications. Science **297**(5582), 787–792 (2002)
6. Coates J (2000) Interpretation of infrared spectra, a practical approach. Encyclopedia of analytical chemistry. 1–23, doi:10.1002/9780470027318.a5606
7. Das, R., Ali, M.E., Abd Hamid, S.B., Ramakrishna, S., Chowdhury, Z.Z.: Carbon nanotube membranes for water purification: a bright future in water desalination. Desalination **336**, 97–109 (2014). doi:10.1016/j.desal.2013.12.026
8. Datsyuk, V., Kalyva, M., Papagelis, K., Parthenios, J., Tasis, D., Siokou, A., Kallitsis, I., Galiotis, C.: Chemical oxidation of multiwalled carbon nanotubes. Carbon **46**(6), 833–840 (2008)
9. Dillon, A.C., Gennett, T., Jones, K.M., Alleman, J.L., Parilla, P.A., Heben, M.J.: A simple and complete purification of single-walled carbon nanotube materials. Adv. Mater. **11**(16), 1354–1358 (1999)
10. Eckmann, A., Felten, A., Mishchenko, A., Britnell, L., Krupke, R., Novoselov, K.S., Casiraghi, C.: Probing the nature of defects in graphene by Raman spectroscopy. Nano Lett. **12**(8), 3925–3930 (2012)
11. Eklund, P., Subbaswamy, K.: Analysis of breit-wigner line shapes in the Raman spectra of graphite intercalation compounds. Phy. Rev. B **20**(12), 5157 (1979)
12. Esteve, W., Budzinski, H., Villenave, E.: Relative rate constants for the heterogeneous reactions of OH, NO_2 and NO radicals with polycyclic aromatic hydrocarbons adsorbed on carbonaceous particles. Part 1: PAHs adsorbed on 1–2 μm calibrated graphite particles. Atmos. Environ. **38**(35), 6063–6072 (2004)
13. Fang, H.-T., Liu, C.-G., Liu, C., Li, F., Liu, M., Cheng, H.-M.: Purification of single-wall carbon nanotubes by electrochemical oxidation. Chem. Mater. **16**(26), 5744–5750 (2004). doi:10.1021/cm035263h
14. Feng, Y., Zhang, H., Hou, Y., McNicholas, T.P., Yuan, D., Yang, S., Ding, L., Feng, W., Liu, J.: Room temperature purification of few-walled carbon nanotubes with high yield. ACS Nano **2**(8), 1634–1638 (2008)

15. Harutyunyan, A.R., Pradhan, B.K., Chang, J., Chen, G., Eklund, P.C.: Purification of single-wall carbon nanotubes by selective microwave heating of catalyst particles. J. Phys. Chem. B **106**(34), 8671–8675 (2002)
16. Hone, J., Batlogg, B., Benes, Z., Johnson, A., Fischer, J.: Quantized phonon spectrum of single-wall carbon nanotubes. Science **289**(5485), 1730–1733 (2000)
17. Hou, P.-X., Liu, C., Cheng, H.-M.: Purification of carbon nanotubes. Carbon **46**(15), 2003–2025 (2008)
18. Hou, P., Liu, C., Tong, Y., Xu, S., Liu, M., Cheng, H.: Purification of single-walled carbon nanotubes synthesized by the hydrogen arc-discharge method. J. Mater. Res. **16**(09), 2526–2529 (2001)
19. Iijima, S.: Helical microtubules of graphitic carbon. Nature **354**(6348), 56–58 (1991)
20. Karatepe, N., Yuca, N.: Hydrogen adsorption on carbon nanotubes purified by different methods. Int. J. Hydrogen Energy **36**(17), 11467–11473 (2011). doi:10.1016/j.ijhydene.2011.01.128
21. Kim, U.J., Furtado, C.A., Liu, X., Chen, G., Eklund, P.C.: Raman and IR spectroscopy of chemically processed single-walled carbon nanotubes. J. Am. Chem. Soc. **127**(44), 15437–15445 (2005)
22. Kobayashi, Y., Nakashima, H., Takagi, D., Homma, Y.: CVD growth of single-walled carbon nanotubes using size-controlled nanoparticle catalyst. Thin Solid Films **464**, 286–289 (2004)
23. Koehne, J.E., Chen, H., Cassell, A.M., Ye, Q., Han, J., Meyyappan, M., Li, J.: Miniaturized multiplex label-free electronic chip for rapid nucleic acid analysis based on carbon nanotube nanoelectrode arrays. Clin. Chem. **50**(10), 1886–1893 (2004)
24. Kong, J., Franklin, N.R., Zhou, C., Chapline, M.G., Peng, S., Cho, K., Dai, H.: Nanotube molecular wires as chemical sensors. Science **287**(5453), 622–625 (2000)
25. Liu, J., Rinzler, A.G., Dai, H., Hafner, J.H., Bradley, R.K., Boul, P.J., Lu, A., Iverson, T., Shelimov, K., Huffman, C.B., Rodriguez-Macias, F., Shon, Y.S., Lee, T.R., Colbert, D.T., Smalley, R.E.: Fullerene pipes. Science **280**(5367), 1253–1256 (1998)
26. Liu, Z., Xiao, B., Wang, W., Ma, Z.: Tensile strength and electrical conductivity of carbon nanotube reinforced aluminum matrix composites fabricated by powder metallurgy combined with friction stir processing. J. Mater. Sci. Technol. **30**(7), 649–655 (2014)
27. Ma, P-C., Kim, J-K.: Carbon nanotubes for polymer reinforcement. CRC Press, Boca Raton (2011)
28. Miyauchi, M., Miao, J., Simmons, T.J., Lee, J.-W., Doherty, T.V., Dordick, J.S., Linhardt, R. J.: Conductive cable fibers with insulating surface prepared by coaxial electrospinning of multiwalled nanotubes and cellulose. Biomacromol **11**(9), 2440–2445 (2010)
29. Naseh, M.V., Khodadadi, A., Mortazavi, Y., Sahraei, O.A., Pourfayaz, F., Sedghi, S.M.: Functionalization of carbon nanotubes using nitric acid oxidation and DBD plasma. World Acad. Sci. Eng. Technol. **49**, 177–179 (2009)
30. Nyquist, R.A.: Interpreting infrared, Raman, and nuclear magnetic resonance spectra. Academic Press, London (2001)
31. Parveen, S., Rana, S., Fangueiro, R.: A review on nanomaterial dispersion, microstructure, and mechanical properties of carbon nanotube and nanofiber reinforced cementitious composites. J. Nanomaterials **2013**, 80 (2013)
32. Peng, Y., Liu, H.: Effects of oxidation by hydrogen peroxide on the structures of multiwalled carbon nanotubes. Ind. Eng. Chem. Res. **45**(19), 6483–6488 (2006)
33. Prasek, J., Drbohlavova, J., Chomoucka, J., Hubalek, J., Jasek, O., Adam, V., Kizek, R.: Methods for carbon nanotubes synthesis-review. J. Mater. Chem. **21**(40), 15872–15884 (2011). doi:10.1039/c1jm12254a
34. Ren, W., Li, F., Chen, J., Bai, S., Cheng, H.-M.: Morphology, diameter distribution and Raman scattering measurements of double-walled carbon nanotubes synthesized by catalytic decomposition of methane. Chem. Phys. Lett. **359**(3–4), 196–202 (2002). doi:10.1016/S0009-2614(02)00686-3

35. Rinzler, A., Liu, J., Dai, H., Nikolaev, P., Huffman, C., Rodriguez-Macias, F., Boul, P., Lu, A.H., Heymann, D., Colbert, D.: Large-scale purification of single-wall carbon nanotubes: process, product, and characterization. Appl. Phys. A Mater. Sci. Process. **67**(1), 29–37 (1998)
36. Sahaym, U., Norton, M.G.: Advances in the application of nanotechnology in enabling a 'hydrogen economy'. J. Materials Sci. **43**(16), 5395–5429 (2008)
37. Scheibe, B., Borowiak-Palen, E., Kalenczuk, R.J.: Oxidation and reduction of multiwalled carbon nanotubes—preparation and characterization. Mater. Charact. **61**(2), 185–191 (2010)
38. Shieh, Y.-T., Liu, G.-L., Wu, H.-H., Lee, C.-C.: Effects of polarity and pH on the solubility of acid-treated carbon nanotubes in different media. Carbon **45**(9), 1880–1890 (2007)
39. Shin, Y.-R., Jeon, I.-Y., Baek, J.-B.: Stability of multi-walled carbon nanotubes in commonly used acidic media. Carbon **50**(4), 1465–1476 (2012)
40. Spitalsky, Z., Tasis, D., Papagelis, K., Galiotis, C.: Carbon nanotube–polymer composites: chemistry, processing, mechanical and electrical properties. Prog. Polym. Sci. **35**(3), 357–401 (2010)
41. Suzuki, T., Suhama, K., Zhao, X., Inoue, S., Nishikawa, N., Ando, Y.: Purification of single-wall carbon nanotubes produced by arc plasma jet method. Diamond Relat Mater **16** (4), 1116–1120 (2007)
42. Tans, S.J., Verschueren, A.R., Dekker, C.: Room-temperature transistor based on a single carbon nanotube. Nature **393**(6680), 49–52 (1998)
43. Tasis, D., Tagmatarchis, N., Bianco, A., Prato, M.: Chemistry of carbon nanotubes. Chem. Rev. **106**(3), 1105–1136 (2006)
44. Thostenson, E.T., Li, C., Chou, T.-W.: Nanocomposites in context. Compos. Sci. Technol. **65** (3), 491–516 (2005)
45. Vaisman, L., Marom, G., Wagner, H.D.: Dispersions of surface-modified carbon nanotubes in water-soluble and water-insoluble polymers. Adv. Funct. Mater. **16**(3), 357–363 (2006)
46. Vione, D., Maurino, V., Minero, C., Pelizzetti, E., Harrison, M.A., Olariu, R.-I., Arsene, C.: Photochemical reactions in the tropospheric aqueous phase and on particulate matter. Chem. Soc. Rev. **35**(5), 441–453 (2006)
47. Walling, C.: Fenton's reagent revisited. Acc. Chem. Res. **8**(4), 125–131 (1975). doi:10.1021/ar50088a003
48. Wang, Y., Shan, H., Hauge, R.H., Pasquali, M., Smalley, R.E.: A highly selective, one-pot purification method for single-walled carbon nanotubes. J. Phys. Chem. B **111**(6), 1249–1252 (2007)
49. Wepasnick, K.A., Smith, B.A., Schrote, K.E., Wilson, H.K., Diegelmann, S.R., Fairbrother, D.H.: Surface and structural characterization of multi-walled carbon nanotubes following different oxidative treatments. Carbon **49**(1), 24–36 (2011)
50. Wong, E.W., Sheehan, P.E., Lieber, C.M.: Nanobeam mechanics: elasticity, strength, and toughness of nanorods and nanotubes. Science **277**(5334), 1971–1975 (1997)
51. Zhang, Y., Bai, Y., Yan, B.: Functionalized carbon nanotubes for potential medicinal applications. Drug Discovery Today **15**(11), 428–435 (2010)
52. Zhao, X., Ohkohchi, M., Inoue, S., Suzuki, T., Kadoya, T., Ando, Y.: Large-scale purification of single-wall carbon nanotubes prepared by electric arc discharge. Diamond Relat Mater **15** (4), 1098–1102 (2006)

Chapter 4
Carbon Nanotube Functionalizations

Abstract Hydrophobic CNTs have shown to be aggregated and precipitated in polar solvents. These have made their handling difficult and limited their applications in various fields including water purification technologies, catalysis, polymers, composites, sensors, and optoelectronics. Here, we reported two covalent functionalization schemes for MWCNTs using HNO_3/H_2O_2 mixture and basic $KMnO_4$ solution. HNO_3/H_2O_2 mixture anchored more –C=O and –OH groups on oxidized (O)-MWCNTs which were less soluble in water. In contrast, $KMnO_4$ unzipped the closed-end tips of MWCNT with a higher number of –COOH functionalities. The group (–COOH) was necessary to improve O-MWCNT dispersion and colloidal stability in both water and acetone solvents. We suggested here the –COOH groups were active in neutral (pH 7.1) and more functioning in alkaline aqueous solutions (pH 10.0), but were inactive in acidic media (pH 3.0). Finally, we proposed a mechanism for the solubilization of MWCNTs to interpret the findings. We proved the observations based on XPS, titration, TEM, Raman spectroscopy, TGA and UV/vis spectroscopy.

4.1 Background

Hydrophobic CNTs have shown to trigger its aggregation followed by precipitation in polar solvents [50]. This has hindered the catalytic rates and performances of CNTs in water purification fields [12]. For instance, CNT hydrophobicity could minimize optimum enzyme loading to develop an effective Nanobiohybrid catalyst [54] for water purifications. In addition, hydrophobic CNTs have shown to decrease its surface affinity for biological adsorbates in wastewater treatment plants [28]. Besides, water-soluble CNTs have needed for fabricating composite membranes, which can be used for both brackish and sea water desalinations [11]. Therefore, manipulation of CNT hydrophobicity has remained a challenging task for developing an effective water purification technology.

Different functionalization techniques, such as poly(styrene-co-p-(4-(4′-vinylphenyl)-3-oxabutanol) [24], amino acid [68], poly L-lactic acid [9], H_2O_2 [45],

R. Das, *Nanohybrid Catalyst based on Carbon Nanotube*,
Carbon Nanostructures, DOI 10.1007/978-3-319-58151-4_4

ozone (O_3) [34], H_2SO_4/HNO_3 [52], HNO_3 [4], $(NH_4)_2S_2O_8$ [49] and so on have been adopted for increasing CNT colloidal stability. The treatments have changed nanotubes' surface charges and energy potentials followed by inserting different functionalities which alter their intra and/or inter molecular interactions and surface properties [52]. For example, introducing –COOH groups on CNT surfaces may improve their water solubility and dispersion [50]. Anchoring –COOH group on CNTs is more important to develop an effective CNT-based Nanobiohybrid catalyst [20] for water purifications [12]. Herein –COOH may act as major attachment site to covalently immobilize molecules such as enzymes on its surface [20] via diimide-activated amidation reaction [28]. The negatively charged $–COO^-$ could specifically bind positively charged water pollutants by electrostatic interactions and subsequently removed them from water [63].

In this study, we evaluated systematically HNO_3/H_2O_2 method [47] and developed a basic $KMnO_4$ method ($KMnO_4$ directly dissolved into water) to yield –COOH decorated MWCNTs. We observed HNO_3/H_2O_2 mixture was unable to produce water soluble MWCNTs in contrast to previous observations [47]. Basic $KMnO_4$ treatment showed to anchor a lot of –COOH functionalities on MWCNT surfaces. It showed to improve their water solubility and dispersion as like the acidic ($KMnO_4$ with H_2SO_4) and alkalescent ($KMnO_4$ with NaOH) mediated O-MWCNTs [3, 52]. Finally, we heeded that the ionic strength and pH of the media could have direct effects in the protonation and deprotonation of the –COOH, which might necessary to manipulate O-MWCNT solubility in water.

XPS was performed to mainly measure the carbon and oxygen percentages along with an accurate identification of surface functionalities generated on O-MWCNTs. An autotitrator was used to quantify two major surface functionalities such as acidic (–COOH) and basic (–OH) on MWCNTs. TEM and Raman spectroscopy were executed to trace out MWCNT morphological changes. TGA was carried out to measure nanotube' weight loss at different temperatures. Finally, we proposed a mechanism to explain the solubility profiles of the O-MWCNTs in different media based on the UV/vis spectroscopy and camera observations.

4.2 Literature Review

4.2.1 Is CNT Hydrophobic in Nature?

In general, carbon is hydrophobic unless it has been functionalized with polar moieties. Oxidation is the easiest of such functionalization, making the carbonaceous support much more hydrophilic. But in the absence of such surface functions, CNT is almost purely based on aromatic, nonpolar graphite sheets, so that interaction with extremely polar molecules such as water is very weak.

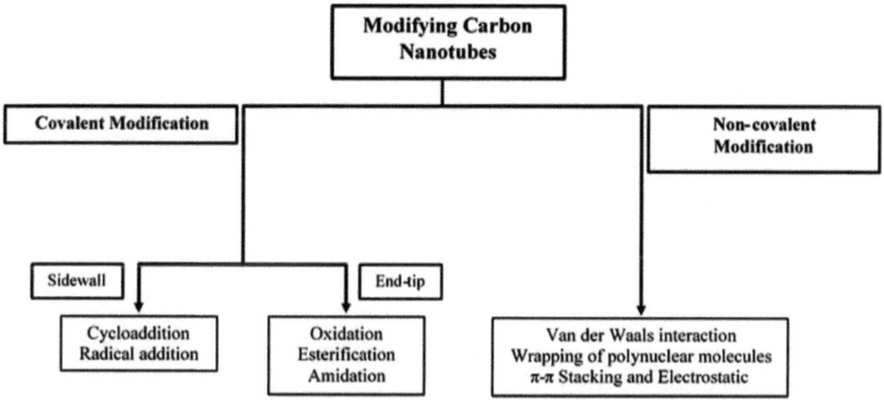

Fig. 4.1 Graphical representation of CNT chemical modifications techniques

4.2.2 *What are the Strategies Commonly Used for CNT Solubilization?*

To enable CNT's purification at individual scale and manipulate their solubility, various covalent and non-covalent functionalization strategies have been adopted [27, 43]. In covalent modification, hydrophilic substituent could be introduced using various wet chemical treatments such as acidic, basic, and oxidizing agents [27]. Both sidewall and tip end functionalizations [10, 55] as shown in Fig. 4.1 have been performed to reduce or eliminate hydrophobicity, increase solubility and colloidal dispersibility [19]. In non-covalent modifications as shown in Fig. 4.1, π–π stacking of polyaromatic compounds and wrapping of polymers have been widely used to increase CNT solubility in different aqueous media [41]. However, covalent modification is preferred over non-covalent surfactant treatments for improving colloidal stability. This is because of three main reasons. First, "tail-to-tail" hydrophobic interactions of surfactants force in non-covalent MWCNT functionalization, which has shown to form nanotube bundles and aggregates [36]. Second, the hydrophobic interactions between CNTs and surfactant molecules are also weaker than covalent bonding due to non-bridged electronic bonds. Finally, surfactant treatments may not be able to completely eliminate CNT impurities. Although many covalent modifications schemes have been proposed in functionalizing MWCNTs; most of them are complicated, involved multiple steps, showed less solubility percentage and difficult purification processes [27].

4.2.2.1 Covalent Modification of CNT

Cycloaddition: Diels–Alder reaction mediated the CNT sidewalls and tip ends functionalization using azide photochemistry is shown in Fig. 4.2. The irradiation

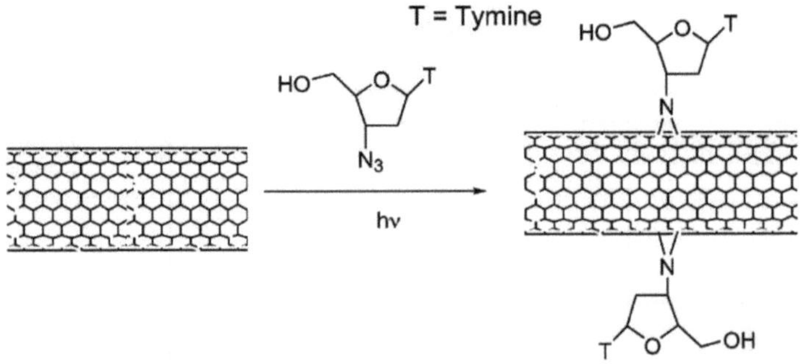

Fig. 4.2 Photo-induced generation of reactive nitrenes in the presence of nanotubes. Figure is adapted with permission from ACS [57]

of the photoactive azidothymidine in the presence of CNTs was found to cause the formation of very reactive nitrene groups in the proximity of the carbon lattice. In a cycloaddition reaction, these nitrene groups couple to the CNT and form aziridine adducts. Another most well-known 1,3 cycloaddition reaction involves azomethine ylides reacting with CNTs, which are of great interest. The addition of a pyrrolidine ring can lead to a variety of functional groups including poly(amidoamine) dendrimers, phthalocyanineaddends, perfluoroalkylsilane groups, and amino ethyleneglycol groups. These groups are second-generation functional moieties which can make CNT surface more reactive with the water molecules for surface reaction.

Radical addition: Molecular dynamics simulation shows the great possibility of CNT sidewall attachments. This can be done using chemical agents such as aryl diazonium, sodium nitrite, ammonium persulfate, etc. However, thermal and photochemical routes have also been found prominent in the successful covalent functionalization of CNT with radicals. Alkyl or aryl peroxides were decomposed thermally and the resulting radicals (phenyl or lauroyl) added to the graphitic network. In an alternative approach, CNT can be heated in the presence of peroxides and alkyl iodides or treated with various sulfoxides, employing Fenton's reagent. The reaction of CNT with succinic or glutaric acid acyl peroxides resulted in the addition of carboxyalkyl radicals onto the sidewalls (Fig. 4.3). This acid-functionalized material can be converted to acid chlorides and then to amides with various terminal diamines.

Oxidation: The methods of CNT impurities oxidation are well corroborated in this chapter. It is effective for introducing the functional groups onto CNT, but during oxidation the C–C bonded networks in CNT are easily broken-down which anchored the functionalities such as COOH, C=O, and OH groups. Rehman et al. [47] studied the solubility profile of MWCNTs using HNO_3/H_2O_2 mixture in different solvents [47]. The group hypothesized the solubility of MWCNTs oxidized by HNO_3/H_2O_2 was lasted for 45 days in water with the formation of –COOH functionalities on MWCNTs. However, the study was not systematic, and most of the results were misinterpreted. For instance, non-quantitative Fourier transform

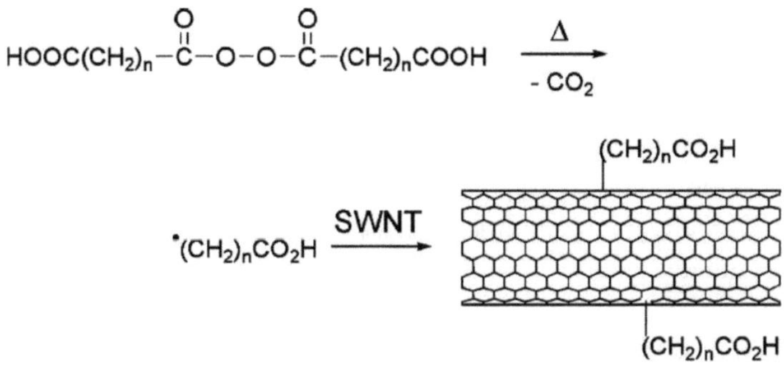

Fig. 4.3 Derivatization reaction with carboxyalkyl radicals by a thermal process. Figure is adapted with permission from ACS [57]

infrared spectroscopy (FTIR) [60] was used to identify the –COOH groups. As a corollary, no conclusion about the quantity of –COOH groups introduced by HNO_3/H_2O_2 treatment can be made. $KMnO_4$ has been popular for CNT functionalization [27]. Although acidic [8, 23, 25, 44, 52, 66] and alkalescent $KMnO_4$ [3, 69] solutions have been used to purify and solubilize CNTs in different solvents, no extensive studies have been adopted using basic $KMnO_4$. Abdul Majid et al. [1] performed MWCNT oxidation using a basic $KMnO_4$ for immobilizing tyrosinase enzyme [1], but unintentionally ignored the solubility tests of O-MWCNTs in aqueous media. In addition, no characterization tool has been used to determine and quantify –COOH functionality generated upon treatment, leaving pitfalls in assessing the validity of the developed method. It should be keep in mind that over oxidation of C–C bond networks may cause the CNT to break-up into small fragments. Among all oxidation methods, H_2O_2 limits the damage of CNT network.

Esterification/Amidation: The COOH groups are used as the anchoring points for most the esterification and amidation reactions. COOH is transformed into an acyl chloride with the use of thionyl or oxalyl chloride which is then reacted with the desired amide, amine, or alcohol. CNT have been deposited on with Ag nanoparticles with the aid of amination reactions. Amide functionalized CNTs have been shown to chelate Ag nanoparticles. CNT modified with acyl chloride react readily with highly branched molecules such as poly(amindoamine), which acts as a template for Ag^+ and later being reduced by formaldehyde. Amino-modified CNT can be prepared by reacting ethylenediamine with an acyl chloride functionalized CNT. Esterification reactions resulted also in soluble functionalized CNTs as shown in Fig. 4.4.

4.2.3 *Non-covalent Modification of CNT*

Non-covalent CNT modification requires van der Waals forces, wrapping of polymers such as surfactants, biomolecules etc. (physical) and π–π interactions as

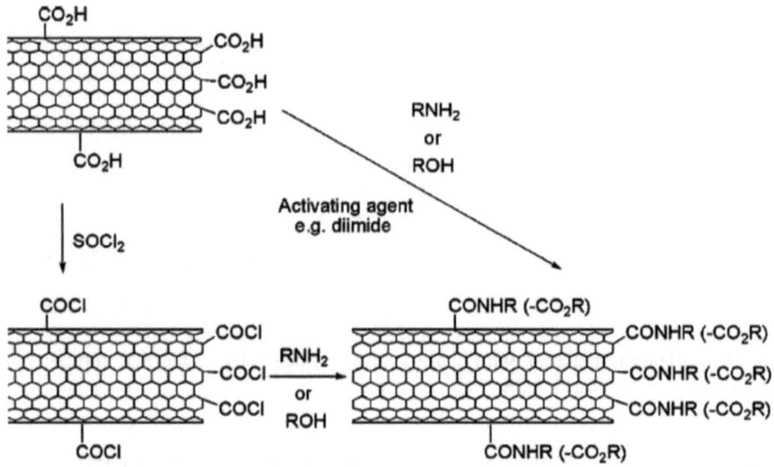

Fig. 4.4 Derivatization reactions of acid-cut nanotubes through the defect sites of the graphitic surface. Figure is adapted with permission from ACS [57]

shown in Fig. 4.1. These interactions preserve the CNT skeleton. There are polynuclear aromatic compounds based moieties, e.g., phenyl, naphthalene, phenanthrene, pyrene and porphyrin systems which have both hydrophilic and/or hydrophobic groups—can be used to solubilize CNT into organic or aqueous phases. The affinity of π–π stacking between the CNT and moieties could determine the rate of CNT dispersibility in water. The greater of π–π stacking (e.g., pyrene amphiphiles) as shown in Fig. 4.5 might have the best solubility than the phenyl amphiphiles with the worse π–π stacking. The polynuclear aromatic compounds can be further modified with NH_2 and COOH groups prior to functionalizing the CNT, if tuning of CNT solubility is required.

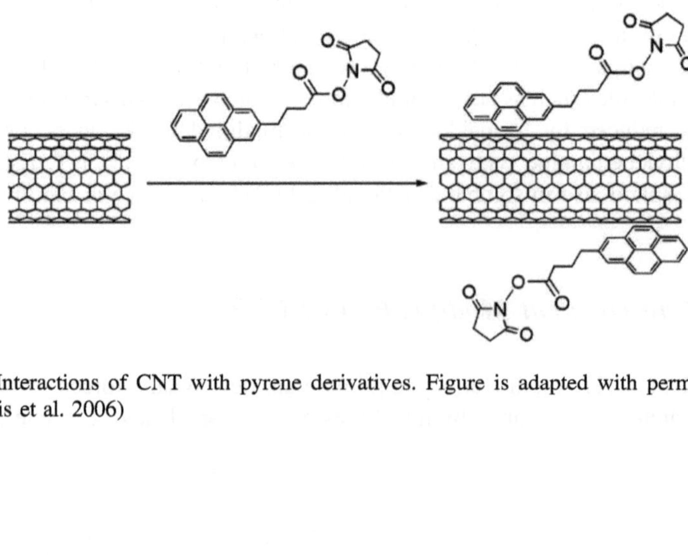

Fig. 4.5 Interactions of CNT with pyrene derivatives. Figure is adapted with permission from ACS (Tasis et al. 2006)

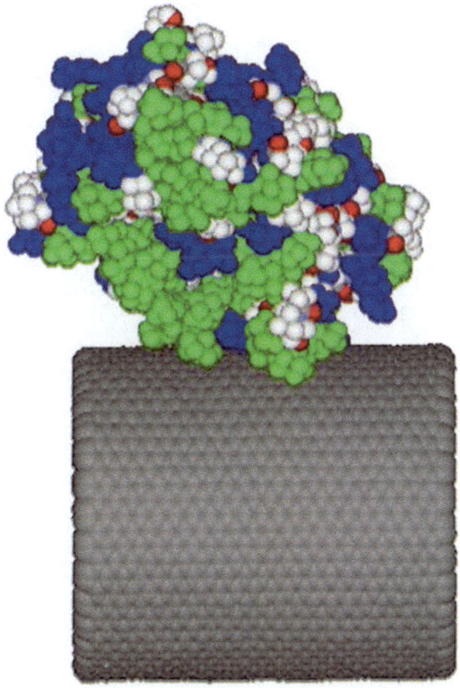

Fig. 4.6 Schematic illustration of lipase adsorbed on CNT by molecular dynamics simulation. The hydrophobic parts are in *green*, hydrophilic parts in *blue*. Figure is adapted with permission from Elsevier [18]

Second, some biomolecules such as proteins/enzymes, polysaccharides, nucleic acids (DNA and RNA), and polymers have been widely studied because of their potential to be used in biological applications. The modification of the CNT with these biomolecules is built with the bottom-up technique. Among these biomolecules, enzymes/proteins have high affinity to CNT due to their diversity of amino acids being hydrophobic or hydrophilic characters (Fig. 4.6). Polysaccharides have been successfully been used to modify CNT forming stable hybrids. In order to make CNT more soluble in water, phospholipids such as lysoglycerophospholipids can be used. The tails of phospholipid wraps around the skeleton of CNT, but the double tailed phospholipids did not have the same ability. Finally, molecules that have bifunctionality are used to modify the CNT. One end of the molecule is polyaromatic compound that interacts with the CNT through π–π stacking. The other end of the same molecule has a functional group such as NH_2, COOH, and –SH.

In summary, the chemistry of CNT is a complex subject and is important to be studied intensely. In order to help the potential readers here we summarize different functionalization schemes of CNTs in Fig. 4.7. This shows that different functionalities have added on CNT surfaces through different routes. The controlled functionalization of CNT has not yet been fully achieved. Solubility continues to be an issue, and new purification and characterization techniques are still needed. It is hoped that with the effort carried out in this chapter, we will be able to witness full control of CNT dispersibility in aquatic phases.

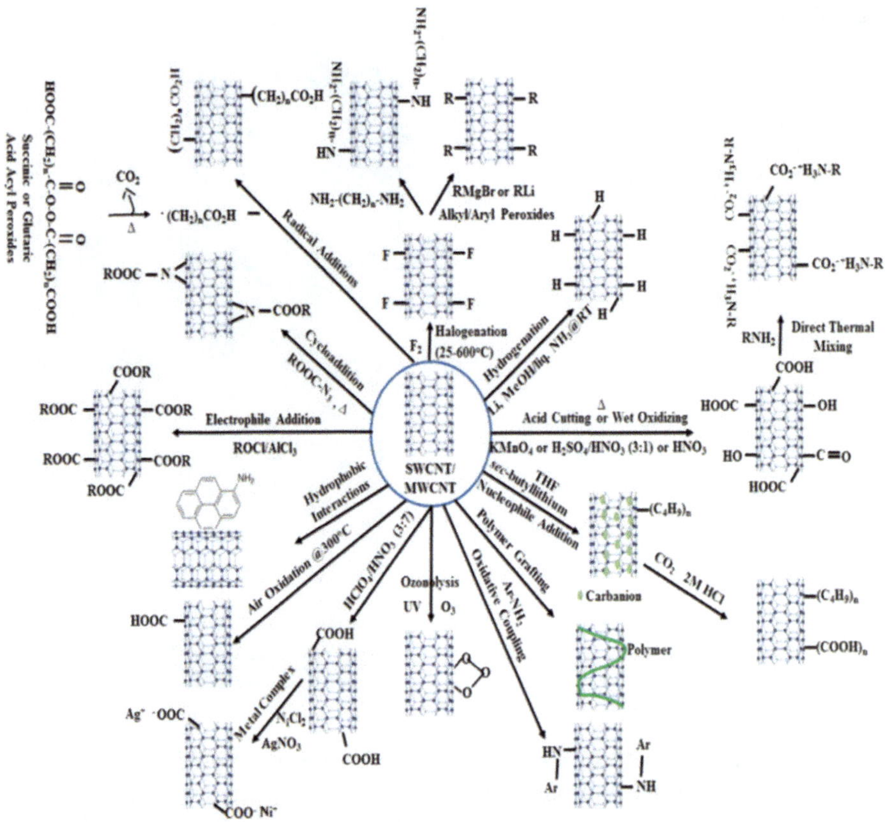

Fig. 4.7 Surface functionalization of CNTs through different routes

4.3 Materials and Methods

4.3.1 Materials and Reagents

Pristine MWCNTs of 13 ± 5 and 4.0 nm in outer, and inner diameters and >1 μm in length were purchased from Bayer MaterialScience AG (Germany). The tubes were prepared by catalytic CVD and contained >95% carbon by weight and were used as received. HNO_3, H_2O_2, and HCl were purchased from Merck Sdn. Bhd. (Malaysia). $KMnO_4$, sodium hydroxide, sodium chloride, ethanol and acetone were purchased from Sigma-Aldrich Sdn. Bhd. (Malaysia). The purity of all reagents was ≥99% except HNO_3 (65%); H_2O_2 (30%); HCl (37%) and ethanol (70%) in water.

4.3.2 *Instrumentation*

XPS was performed by measuring a small quantity (2–3 mg) of MWCNTs and grounded on a copper tape (1 × 1 cm^2). Specimens were then introduced into the XPS system (DAR 400, Serial No-0463, Omicron NanoTechnology, Germany) and conducted using Al Kα X-rays (1486.7 eV) at 10 kV. Obtained spectra were processed using commercially available software (CasaXPS), and atomic concentrations of the elements and each functional group were quantified by integrating of the relevant photoelectron peaks.

An autotitrator was used to measure the titration end point with Mettler Toledo T-50, USA. Morphological characterizations of the MWCNTs were performed using HRTEM (JEOL JEM-2100F, USA) at 200 kV. Raman spectra were acquired for 10 min at a laser power of 100 on Ar^+ laser (514 nm) focused (50X objective) to a spot size of about 1.5–2.0 μm (Renishaw inVia, UK). Spectra were collected at exposure time 10 and laser power 100. TGA (TGA/SDTA 851, Mettler Toledo, USA) was performed under air flow (50 ml) between 25 and 1000 °C at 10 °C/min. Finally, UV/vis spectrophotometer (Shimadzu, UV-1700; Japan) was performed for measuring the CNT suspension's transmittance.

4.3.3 *Wet Chemical Oxidations of MWCNT*

Two different wet chemical oxidizing agent mixtures were used to oxidize pristine MWCNTs as follows:

(a) **HNO_3/H_2O_2 Treatment**: About 0.5 g of pristine MWCNTs was dispersed into a 25 ml mixture (70:30) of HNO_3 (60 wt%) and H_2O_2 (30 wt%), and the mixture was sonicated for 5 h at 50 °C, 40 kHz.

(b) **$KMnO_4$ Treatment**: A solution of 0.1 M $KMnO_4$ was prepared by dissolving 0.79 g of $KMnO_4$ into a 50 ml of deionized water. About 0.5 g of pristine MWCNTs was dispersed into a 50 ml of 0.1 M $KMnO_4$ solution and was sonicated for 5 h at 50 °C, 40 kHz. The mixture was cooled at room temperature (RT) followed by adding 10 ml of concentrated HCl (36 wt%). Finally, the mixture was then stirred for 30 min to remove Mn containing by-products that form insoluble complexes [25].

4.3.3.1 Removal of the Residual Impurities

O-MWCNTs were extracted from the residual acids, oxidizing agents, metallic by-products and carbonaceous impurities, by repeated cycle of dilution and centrifugation (dissolved in 1.0 L of deionized water and centrifuged (Eppendorf-5430R; Germany)) at 7000 rpm for 20 min. The supernatant was carefully collected when the O-MWCNTs were precipitated at the bottom of the polyethylene centrifuge tube. The procedure was repeated 5–7 times until the resistivity of the supernatant was greater than 0.5 MΩ cm and pH was ~7.0. O-MWCNTs were then rinsed with ethanol (70 wt%) and dried overnight in a vacuum oven at 80 °C.

4.3.3.2 Total Acidic and Basic Groups Quantification

Titration was performed to measure total acidic (mainly –COOH) and basic (mainly –OH) groups of both the pristine and O-MWCNTs. About 0.1 g of MWCNTs was added into 50 ml of 0.01 M NaOH and HCl solutions for quantifying total acidic and basic groups on CNTs, respectively. The mixtures were then stirred for 48 h by a magnetic stirrer at RT to equilibrate MWCNTs with the solutions. MWCNTs with NaOH and HCl mixtures were then titrated with 0.01 M HCl and NaOH solutions, respectively. The complete titration methods were performed in an autotitrator which detects titration end point upon completion. Excess NaOH and HCl volumes were used to calculate total –COOH and –OH groups present on MWCNTs by using back titration calculations [31].

4.3.3.3 Stock Solution Preparations and Solubility Measurements

About 1.0 mg of pristine and O-MWCNTs was added separately into 5.0 ml of water (pH 7.1), acetone and water of having acidic (pH 3.0) and basic (pH 10.0) in order to measure their colloidal stability and solubility. Herein pH was adjusted to 3.0 and 10.0 adding HCl and NaOH, respectively. The solutions were then sonicated for 30 min at 40 kHz in order to get stable colloidal suspensions. Within 30 min all MWCNTs were dispersed in such a way that no visible aggregates were observed in both water and acetone solvents. After sonication, undesirable dusts were removed carefully, and every solution was left for one month to see the dispersibility and colloidal stability of both the pristine and O-MWCNTs. All the observations were captured in a digital camera. The transmittance was measured using a UV/vis spectrophotometer.

4.4 Results

4.4.1 Predicted Chemical Reactions of HNO_3/H_2O_2 and $KMnO_4$ with MWCNT

In order to achieve soluble MWCNTs, oxidizing agent such as $KMnO_4$ was found promising compared with HNO_3/H_2O_2 treatment. The methods not only purified MWCNTs, but also oxidized their skeleton and introduced functionalities as shown

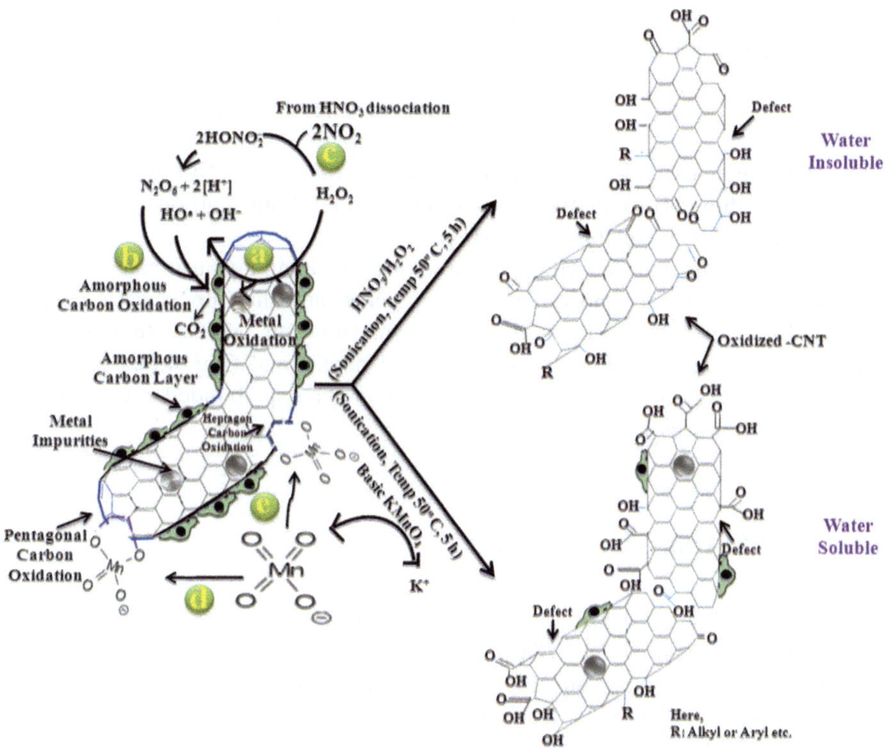

Fig. 4.8 Schematic illustration of the localized catalytic reactions of (**a–c**) HNO_3/H_2O_2 and (**d** and **e**) $KMnO_4$ with pristine MWCNTs

in Fig. 4.8. The method (HNO_3/H_2O_2) can functionalize MWCNTs through different routes. The metals that are usually present in pristine CNTs may act as catalysts to produce hydroxyl radical ($OH^{\cdot}$) through Fenton's chemistry [62] (Fig. 4.8a), which is strong oxidizing agent than H_2O_2 as shown in Eq. 4.1.

$$H_2O_2 + \text{metal (reduced)} \rightarrow OH^{\circ} + OH^- + \text{metal (oxidized)} \tag{4.1}$$

The radical ($OH^{\cdot}$) could then react with amorphous carbon impurities of pristine CNTs [66] and convert them into CO_2 [64] as shown in Eq. 4.2 (Fig. 4.8b).

$$4\,OH^{\cdot} + C \rightarrow CO_2 + 2H_2O \tag{4.2}$$

The strong oxidizing nature of HNO_3 can also be reflected by its large positive reduction potential (E_0^r) as followed in Eq. 4.3.

$$NO_3^- + 2\,H^+ + e^- \rightarrow NO_2 + H_2O, \quad E_0^r = 0.79\,V \tag{4.3}$$

Although H_2O_2 has less affinity to react with HNO_3, H_2O_2 can operate with NO_2 as shown in Eq. 4.4 (Fig. 4.8c).

$$2NO_2 + H_2O_2 \rightarrow 2HONO_2 \rightarrow 2[H^+] + N_2O_6 \tag{4.4}$$

where dinitrogenhexoxide (N_2O_6) is an isomer of two nitrites (NO_3). Because of its weak N–O, this has potentiality to oxidize amorphous carbon impurities of MWCNT to different functionalities (Fig. 4.8b).

Although permanganate (MnO_4^-) is a strong oxidizing agent; it cannot produce any strong free radicals and other etching agents, and the impurities exposed to $KMnO_4$ have less affinity to oxidize by them. However, MnO_4^- treatments can effectively oxidize pentagonal (Fig. 4.8d) and heptagonal (Fig. 4.8e) sp^3 carbon atoms which have shown to oxidize easily rather than sp^2 bonded π-electron graphitic CNT skeletons [45].

4.4.2 XPS Analysis of MWCNT

XPS can offer higher surface sensitivity than FTIR [38] for identifying specific surface functionalities of MWCNTs [13]. Broad scan XPS peaks of pristine and O-MWCNTs are given in Fig. 4.9. Figure 4.10 shows XPS results of high-resolution C1s and O1s deconvoluted different peak components of both the pristine and O-MWCNTs. In addition, Table 4.1 represents the elemental ratio and relative percentages of each peak component. According to Fig. 4.10, presence of aliphatic C–C at 284.9 eV (pristine and MWCNT-HNO_3/H_2O_2), aromatic C=C at 284.3 eV (MWCNT-$KMnO_4$), C–O–C at 286.5 eV (pristine), C=O at 287.9 eV (MWCNT-HNO_3/H_2O_2), C–O at 285.7 eV (MWCNT-$KMnO_4$) and O–C=O at

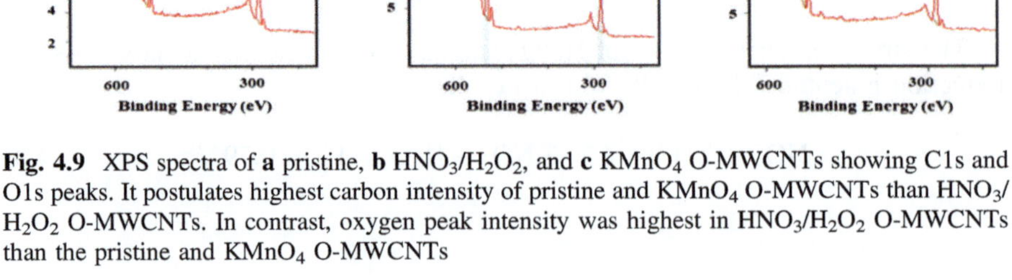

Fig. 4.9 XPS spectra of **a** pristine, **b** HNO_3/H_2O_2, and **c** $KMnO_4$ O-MWCNTs showing C1s and O1s peaks. It postulates highest carbon intensity of pristine and $KMnO_4$ O-MWCNTs than HNO_3/H_2O_2 O-MWCNTs. In contrast, oxygen peak intensity was highest in HNO_3/H_2O_2 O-MWCNTs than the pristine and $KMnO_4$ O-MWCNTs

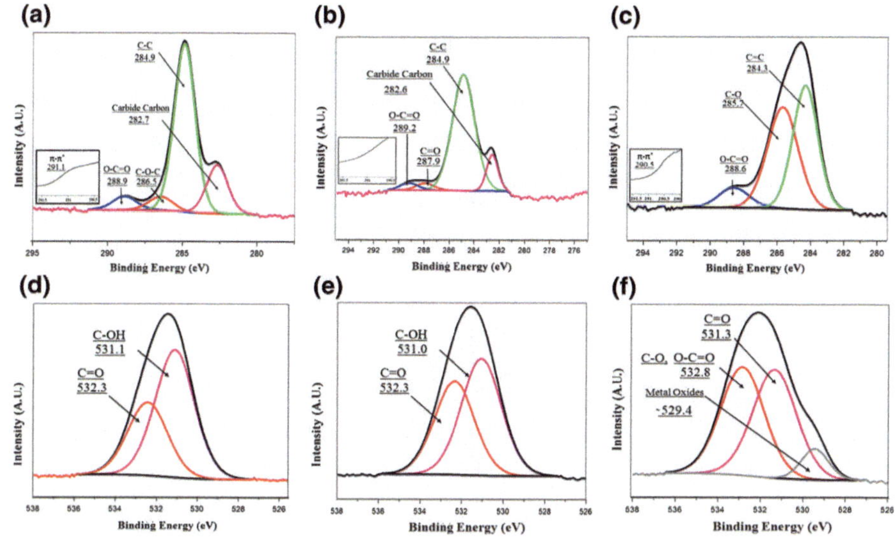

Fig. 4.10 High resolution (**a**–**c**) C1s and (**d**–**f**) O1s XPS spectra of (**a** and **d**) pristine, (**b** and **e**) HNO_3/H_2O_2 and (**c** and **f**) $KMnO_4$ O-MWCNTs, respectively

288.9 (pristine), 289.2 (MWCNT-HNO_3/H_2O_2), and 288.6 eV (MWCNT-$KMnO_4$) in C1s peak (Fig. 4.10a–c) [30, 37, 51, 67]. The O1s spectra deconvoluted into C–OH at 531.1 (pristine) and 531.0 eV (MWCNT-HNO_3/H_2O_2) [22, 39], C=O at 532.3 (pristine and MWCNT-HNO_3/H_2O_2) [2, 70], and 531.3 eV (MWCNT-$KMnO_4$) [25] and a peak at 532.8 eV for either O–C=O [48, 58] or C–O [53] (MWCNT-$KMnO_4$) (Fig. 4.10d–f). In addition, two additional peaks at 282.7 (pristine) and 282.6 eV (MWCNT-HNO_3/H_2O_2) were assigned for the presence of some carbide carbons (Fig. 4.10a, b) [6]. Pristine MWCNT has generic carbide particles which could produce during its synthesis by saturating the catalysts and carbon fragments. For instance, since it contains metal impurities such as Al could be existed as Al_4C_3. The carbide was not eliminated by HNO_3/H_2O_2, but oxidized to form metal oxide by $KMnO_4$ O-MWCNT at 529.4 eV (Fig. 4.10f) [33].

4.4.3 Titration Analysis

Table 4.2 represents mmol of total –COOH and –OH functionalities as a function of wet oxidizing treatments (details are given at Fig. 4.11). This shows $KMnO_4$ treatment produced more –COOH (4.0 mmol/g), whereas HNO_3/H_2O_2 anchored more –OH (1.8 mmol/g) groups on O-MWCNT surfaces.

Table 4.1 XPS analyses of pristine and O-MWCNTs

Samples	Elemental composition (at. %)		Element ratio	Concentration of chemical species (%)										
				C1s							O1s			
	C	O	O/C	C–C	C=C	C–O–C	C=O	C–O	O–C=O	Carbidic carbon	C–OH	C=O	O=C=O/C–O	Metal Oxide
Pristine	68.80	31.20	0.45	67.8	–	6.32	–	–	6.31	19.57	63.61	36.39	–	–
MWCNT-HNO_3/H_2O_2	51.58	48.42	0.94	79.9	–	–	3.81	–	3.81	12.52	55.52	44.48	–	–
MWCNT-$KMnO_4$	63.47	36.53	0.58	–	45.51	–	–	45.47	9.02	–	–	45.85	45.79	8.36

Table 4.2 Concentration of –COOH and –OH groups on pristine and O-MWCNT surfaces

Treatment conditions	–COOH contents (mmol/g)	–OH Contents (mmol/g)
Pristine MWCNT	3.1	1.1
MWCNT-HNO_3/H_2O_2	3.3	1.8
MWCNT-$KMnO_4$	4.0	1.5

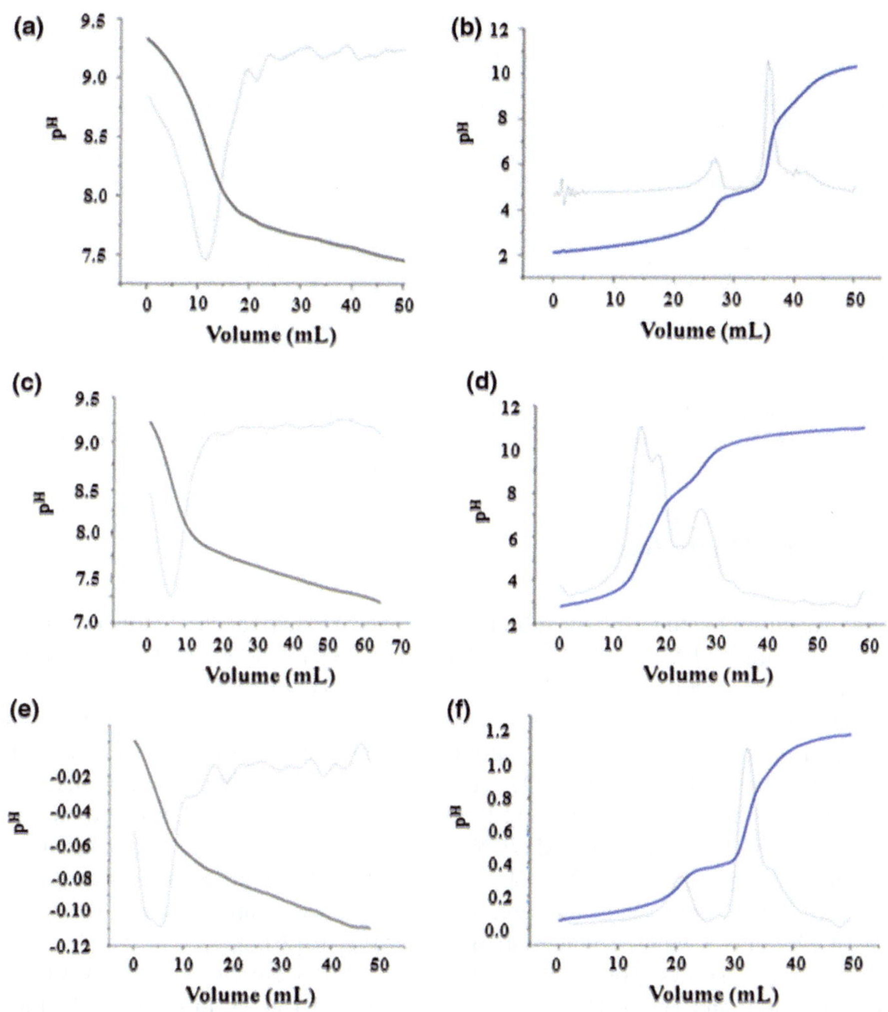

Fig. 4.11 Acid (**a**, **c**, and **e**) and base (**b**, **d**, and **f**) titration curves of (**a** and **b**) pristine, (**c** and **d**) HNO_3/H_2O_2, and (**e** and **f**) $KMnO_4$ O-MWCNTs, respectively. It shows less acidic (mainly –COOH) and basic (mainly –OH) groups present on pristine MWCNTs (**a** and **b**). HNO_3/H_2O_2 O-MWCNTs depict more –OH (**d**), whereas $KMnO_4$ O-MWCNT shows more –COOH groups (**e**)

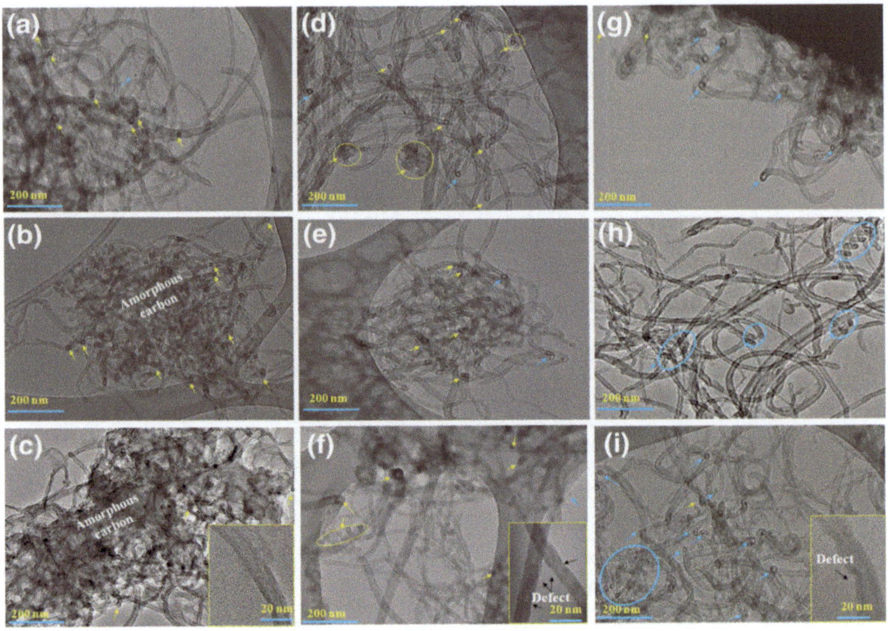

Fig. 4.12 Low magnification TEM images of (**a–c**) pristine; (**d–f**) HNO_3/H_2O_2 and (**g–i**) $KMnO_4$ O-MWCNTs. Defects are indicated by *black arrows* (insert of **c**, **f**, and **i**). *Yellow arrows* and circles indicate close, whereas *blue circles* and arrows depict open MWCNT tip ends

4.4.4 TEM Analysis

TEM was used to study the overall effects of two oxidative agent treatments on the lattice structures of MWCNT. Figures 4.12 and 4.13 show low- and high-magnified TEM images of pristine MWCNTs (a, b and c), HNO_3/H_2O_2 (d, e and f) and $KMnO_4$ O-MWCNTs (g, h, and i), respectively. Pristine MWCNTs reflected clumped, cloudy and amorphous carbon containing MWCNTs (Fig. 4.12a–c). In contrasts, both the HNO_3/H_2O_2 (Fig. 4.12d–f) and $KMnO_4$ (Fig. 4.12g–i) O-MWCNTs appeared in clean and loosely connected bundles. Long and straight tubular sidewalls of pristine MWCNTs indicated defects free or intact lattice structure (insert of Fig. 4.12c). However, MWCNTs oxidized by HNO_3/H_2O_2 showed a few defects in the outermost nanotubes' graphite sheet (insert of Fig. 4.12f) than $KMnO_4$ counterpart (insert of Fig. 4.12i).

Figure 4.13 shows most of the MWCNT tip ends were closed in both pristine (Fig. 4.13a–c) and HNO_3/H_2O_2 O-MWCNTs (Fig. 4.13d–f). In contrast, $KMnO_4$ completely unzipped most of the MWCNT closed-end tips (Fig. 4.13g–i).

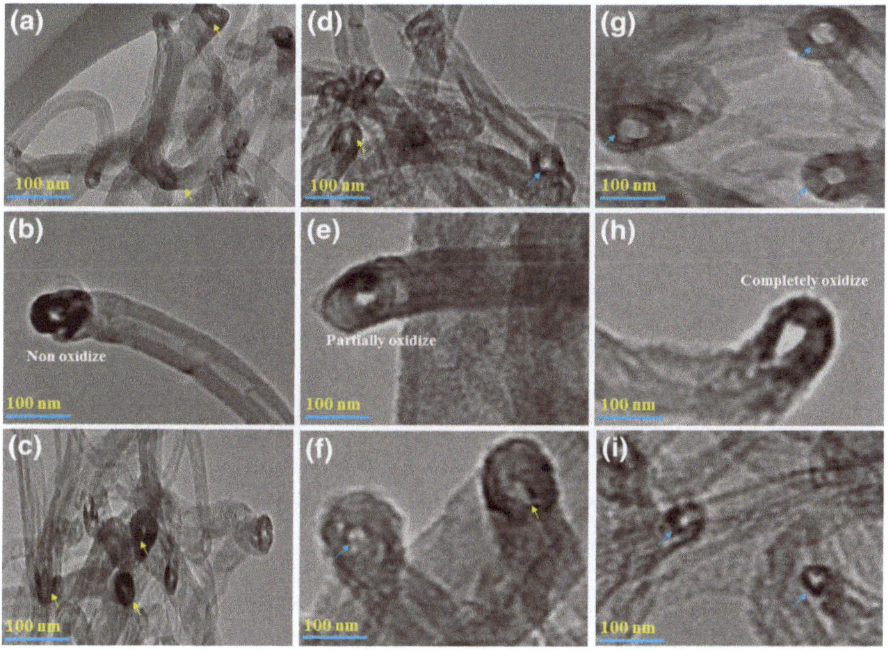

Fig. 4.13 High magnification TEM images of (**a–c**) pristine; (**d–f**) HNO_3/H_2O_2 and (**g–i**) $KMnO_4$ O-MWCNTs. It clearly demonstrates HNO_3/H_2O_2 can partially oxidize the tips of MWCNT, whereas complete oxidation followed by tips opening was possible by $KMnO_4$ treatment. *Yellow arrows* indicate close, whereas blue arrows depict open MWCNT tip ends

4.4.5 Raman Spectroscopy Analysis

Raman spectroscopy was performed to measure the intensity of defects and the quality of MWCNT graphite sheets before and after oxidation processes (Fig. 4.14). Typically, MWCNT represents two significant high frequency bands called D and G-bands near 1330 and 1585 cm^{-1}, reflecting structural disorder and graphite in-plane vibrations, respectively [14, 16]. In this study, D- and G-bands were appeared at ~1349 and ~1588 cm^{-1} both in pristine and O-MWCNTs, respectively. In addition, owing to moderate defect density, D' band was observed for only HNO_3/H_2O_2 O-MWCNTs at ~1615 cm^{-1} [15, 42]. Figure 4.14 shows D-band intensities were practically constant in both the pristine and O-MWCNTs, while G- and D'-band intensities were moderately increased specially for HNO_3/H_2O_2 O-MWCNTs. Furthermore, the intensity ratio (I_D/I_G) represents the actual quantitative measure of defects, purity and distribution of substituent of CNTs [7, 68]. Lowest intensity ratio of the I_D/I_G (1.32) of MWCNT-HNO_3/H_2O_2 was observed as compared with pristine (1.46) and $KMnO_4$ (1.42) O-MWCNTs (Fig. 4.14). The intensity ratio of the D and D' peaks ($I_D/I_{D'}$) can also be used for

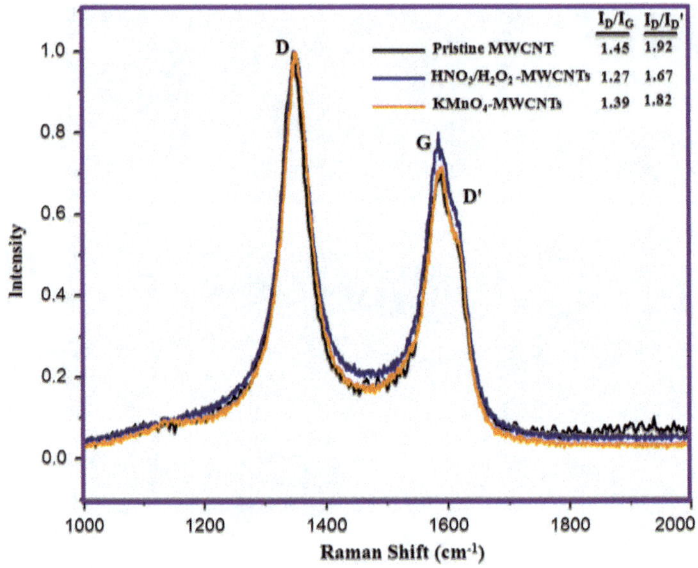

Fig. 4.14 Normalized Raman spectra of pristine and O-MWCNTs

distinguishing the defects associated with sp^2 and sp^3 carbon atoms [15]. Herein ($I_D/I_{D'}$) was highest for pristine (1.92) and $KMnO_4$ (1.82) than HNO_3/H_2O_2 (1.67) O-MWCNTs.

4.4.6 TGA Analysis

TGA was performed to measure the oxidation defects and overall quality of before and after the treatments of MWCNTs using HNO_3/H_2O_2 and $KMnO_4$. CNTs have shown different oxidation behaviors under different oxidative steps [13, 51]. Figure 4.15 shows TGA of pristine and O-MWCNTs, and their DTG spectra. Herein oxidation temperature means the temperature at which MWCNTs lose their weight and produce the highest derivative weight curve. At first, pristine and $KMnO_4$ O-MWCNTs showed lowest decomposition temperatures at around 100 °C, which can be attributed to the pyrolytic evolution of hydroxyl and/or water molecules [13]. TGA of $KMnO_4$ O-MWCNTs showed second decomposition temperature at between 200 and 450 °C compared with pristine and HNO_3/H_2O_2 O-MWCNTs. This was due to the decarboxylation of carboxyl functionalities and elimination of –OH groups present on $KMnO_4$ O-MWCNTs [21, 56]. However, highest decomposition temperatures were found similar for all MWCNTs at around 700 °C [13].

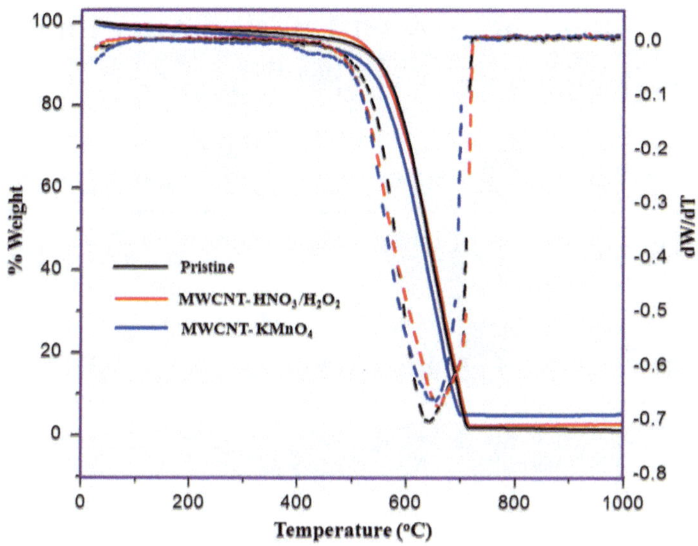

Fig. 4.15 TGA (*solid lines*) and DTG (*dashed lines*) curves of the MWCNTs before and after oxidative treatments

4.4.7 Colloidal Stability of MWCNT

Figure 4.16 shows the colloidal stability of pristine and O-MWCNTs as a function of time. Herein panels I, II, III, and IV represent MWCNT dispersibility in neutral (7.1), acidic (3.0), and basic (10.0) water pH; and acetone, respectively. The acidic and basic solvents were selected because of their established roles in MWCNT solubility [50].

According to (Fig. 4.16 (Panel I)), hydrophobic pristine MWCNTs were not well dissolved in water (a) prior to the oxidation treatments. But O-MWCNTs demonstrated different degrees of solubility depending on the oxidation states and functionalities present over time. MWCNTs oxidized with HNO_3/H_2O_2 were found to be precipitated after a week (b), whereas $KMnO_4$ O-MWCNTs showed good colloidal stability (c).

MWCNTs solubility was shown to decrease at pH <3.0 and >10.0 [50]. As shown in Fig. 4.16 (Panel II), temporary water solubility of pristine MWCNTs was observed (a). Although HNO_3/H_2O_2 O-MWCNTs were not reactive (b), $KMnO_4$ O-MWCNTs showed considerable solubility only for an hour and then precipitated rapidly (c). The result was fully consistent with the previous studies [50, 67].

Figure 4.16 (Panel III) shows the colloidal stability of MWCNTs could be increased at alkaline pH 10.0. MWCNTs oxidized by HNO_3/H_2O_2 showed good dispersion only for an hour (b), then rapidly inactivated and precipitated at the bottom of the vial after one week. However, $KMnO_4$ O-MWCNTs produced a homogenous blackish solution which was lasted for more than one month (c).

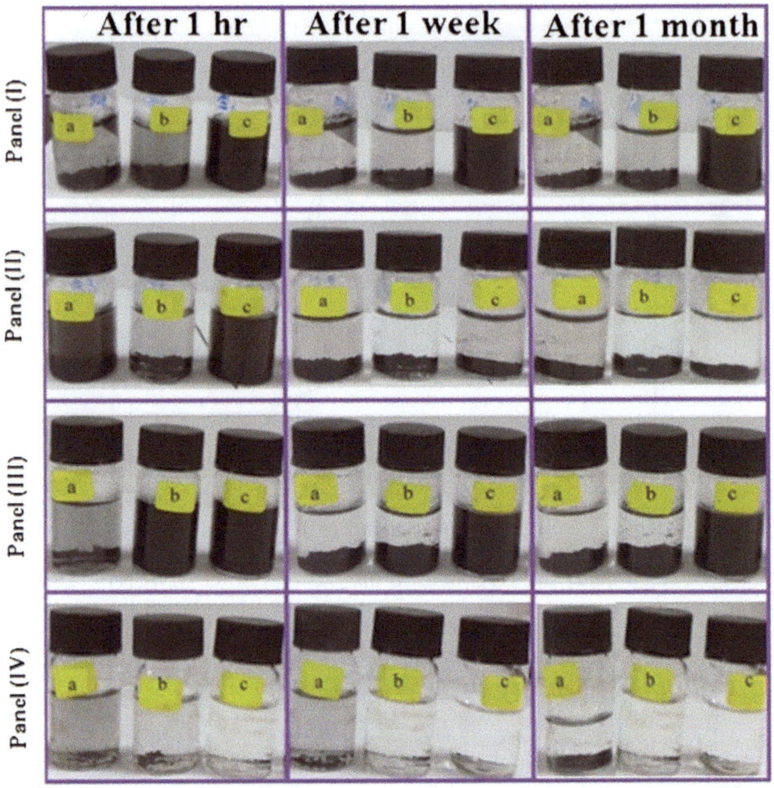

Fig. 4.16 The solubility profiles of MWCNT in water of (panel I) pH 7.1, (panel II) 3.0, (panel III) 10.0; and (panel IV) acetone containing 1.0 mg of MWCNTs in 5.0 mL solutions. Pictures were captured at one hour, one week and one month intervals after the solutions were sonicated for 30 min. Herein *a*, *b* and *c* represent pristine, HNO_3/H_2O_2 and $KMnO_4$ O-MWCNTs, respectively

Besides testing water solubility, MWCNTs were also dispersed into acetone solvent as shown in Fig. 4.16 (Panel IV). Pristine MWCNTs were moderately soluble in acetone (a). However, the effect was not sustained for long time, since it precipitated after one week and one month. MWCNTs oxidized with HNO_3/H_2O_2 (b) and $KMnO_4$ (c) were insoluble in acetone solvent.

All visual observations of the solubility profiles of pristine and O-MWCNTs were verified using UV/vis spectroscopy in order to measure the quantitative transparency of all solutions (Fig. 4.17) [46]. It was shown that the visual observations were similar to the quantitative evaluation of UV/vis spectroscopy. MWCNTs with greater solubility showed increased transmittance than those were coagulated and precipitated, supporting the visual findings. Finally, in order to settle the solubility roles of –COOH group in pH 10.0, we added 1.0 mg of pristine and O-MWCNTs into 5.0 ml of 10 wt/vol.% NaCl solutions and sonicated for 30 min as shown in Fig. 4.18. Pristine MWCNTs were precipitated rapidly

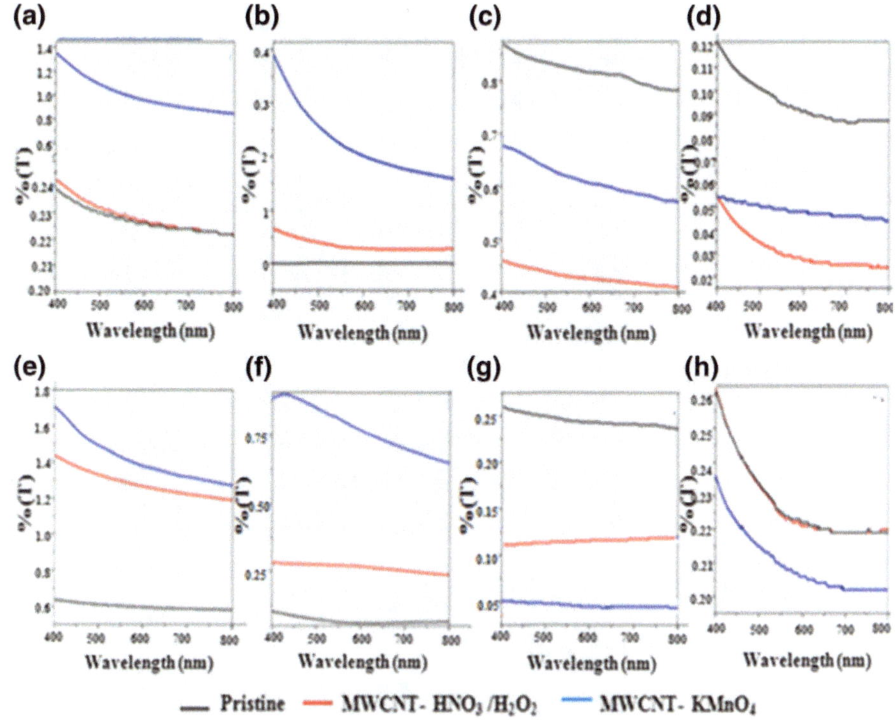

Fig. 4.17 UV–Vis spectra of MWCNTs in water of pH (**a** and **b**) 7.1; (**c** and **d**) 3.0; (**e** and **f**) 10.0; and (**g** and **h**) acetone, which measured after an hour (**a**, **c**, **e**, and **g**) and one month (**b**, **d**, **f**, and **h**), respectively

(Fig. 4.18 (Panel I)). $KMnO_4$ O-MWCNTs were precipitated completely and produced a clean solution (Fig. 4.18 (Panel III)) compared with HNO_3/H_2O_2 (Fig. 4.18 (Panel II)) O-MWCNTs from solvent with increasing time.

4.5 Discussion

Solubilization of MWCNTs was developed using two oxidative agent treatments such as HNO_3/H_2O_2 and $KMnO_4$ in this study. According to Table 4.1, oxygen to carbon (O/C) ratio was increased from 0.45 of pristine (Fig. 4.9a) to 0.94 and 0.58 of MWCNTs oxidized with HNO_3/H_2O_2 (Fig. 4.9b) and $KMnO_4$ (Fig. 4.9c), respectively. It indicates HNO_3/H_2O_2 and $KMnO_4$ were successful to oxidize amorphous carbon impurities followed by grafting of different oxygen functionalities on O-MWCNTs. Two folds increased (O/C) ratio of HNO_3/H_2O_2 might be due to its ability to graft higher amounts of C=O and –OH groups upon amorphous carbon oxidation of MWCNTs. Absence of a C1s peak at 290.5 eV (insert of Fig. 4.10b) postulated the increased disruption of π-electron system of CNT sidewall [66],

Panel (I)

Panel (II)

Panel (III)

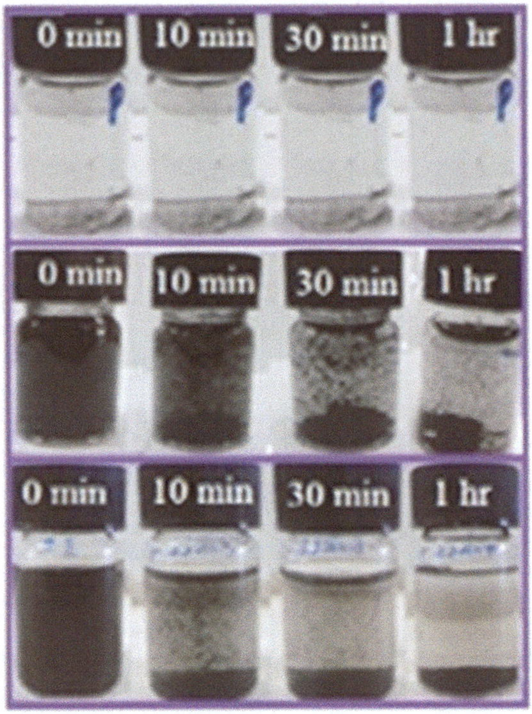

Fig. 4.18 Effects of NaCl solutions on the solubility of (Panel I) pristine, (Panel II) HNO_3/H_2O_2 and (Panel III) $KMnO_4$ O-MWCNTs at pH 10.0. Images were captured at 0, 10 and 30 min and 1 h incubations at 25 ± 2 °C

which is supposed to be present in the intact CNT graphite layers. As a result, we hypothesized that the functionalities that were generated on HNO_3/H_2O_2 O-MWCNTs due to the oxidation of amorphous carbons presence on the sidewall graphitic skeletons of MWCNTs. On the other hand, a peak at 290.5 eV in $KMnO_4$ O-MWCNTs (insert of Fig. 4.10c) suggested less-defected sidewall structure (moderate disruption of delocalized π electrons), which indicated functionalities might be incorporated into the CNT tip ends followed by localized cap oxidation (unzipping).

Pristine MWCNTs showed generic functionalities such as (C–O–C and O–C=O), and (C–OH and C=O) in C1s and O1s, respectively (Fig. 4.10a, d). The groups were probably generated unintentionally when MWCNTs were exposed to the environment containing natural oxidizing agents such as ozone and hydroxyl radicals [17, 61]. The higher oxygen percentage (48.42 at.%) of HNO_3/H_2O_2 O-MWCNTs (Fig. 4.9b) was approved from the presence of C=O and O–C=O in C1s (Fig. 4.10b), whereas higher fractions might come from C–OH and C=O in O1s (Fig. 4.10e and Table 4.1).

Although oxygen percentage of $KMnO_4$ O-MWCNTs (36.53 at.%) (Fig. 4.9c) was decreased than the HNO_3/H_2O_2 O-MWCNTs (Fig. 4.9b), most of the fractions were originated from O–C=O and C=O groups (Table 4.1). A peak at 532.8.0 eV in O1s spectra confirmed a large fractions of oxygen atoms was present as either O–C=O [47, 56] or C–O [53] groups in $KMnO_4$ O-MWCNTs (Fig. 4.10f). Further, notable owing to oxygen (C=O) groups in O1s was generated from O–C=O linked to aromatic rings in $KMnO_4$ O-MWCNTs (Fig. 4.10f) [26]. Therefore, O–C=O surface species were the major final oxidation products in $KMnO_4$ O-MWCNTs. Simulation study reflects that a MWCNT with 15 nm in diameter and 1 μm in length is composed of nine walls consisting of about 20 million carbon atoms. When all closed ends are oxidized, total number of exposed carbon atom becomes 3000 out of the 20 million [66]. These hugely opened carbon atoms (–C) at tips can be easily oxidized by $KMnO_4$ to form –COOH because of their higher reactivity. The finding is analogous with HRTEM analyses as shown in Fig. 4.13g–i.

Although constant D-band intensities were observed for both the pristine and O-MWCNTs in Raman spectra (Fig. 4.14), appearance of D'-band in HNO_3/H_2O_2 O-MWCNTs was suggested moderate defect vacancies on O-MWCNTs [15, 32] compared with pristine and $KMnO_4$ O-MWCNTs (Fig. 4.14). Defects could be generated as various natures such as edges, vacancies, boundaries, implanted atoms as well as defects in graphitic sp^2 and amorphous sp^3 carbon hybridizations [15]. However, increasing G-band intensity of HNO_3/H_2O_2 O-MWCNTs suggested effective removal of amorphous carbons from the MWCNTs [66] compared with pristine and $KMnO_4$ O-MWCNTs (Fig. 4.14). In order to analyze D and G- bands' intensities, we further interpreted I_D/I_G ratios of the pristine and O-MWCNTs (Fig. 4.14). A higher I_D/I_G (1.39) of $KMnO_4$ O-MWCNTs was found than HNO_3/H_2O_2O-MWCNTs (1.27), suggesting attachment of different functionalities on nanotube surfaces [58]. The results were consistent with the earlier findings for sidewall functionalization [65] and distribution of the substituent [68]. However, a higher I_D/I_G ratio (1.45) of pristine MWCNTs may be due to the presence of sp^3-hybridized non-nanotube carbon impurities such as amorphous carbons [5]. Another ratio of ($I_D/I_{D'}$) was highest for pristine (1.92) and $KMnO_4$ O-MWCNTs (1.82) compared with HNO_3/H_2O_2 O-MWCNTs (1.67) (Fig. 4.14). The ratio was maximum for sp^3, and it could decrease for vacancy-like defects and reach a minimum for boundaries in graphite [15]. Consequently, we can conclude that the defects generated by the HNO_3/H_2O_2 treatment were associated with graphitic carbon network, whereas defects in $KMnO_4$ O-MWCNTs may originate from sp^3 carbon atoms. This supports our earlier XPS observations discussed in Sect. 4.4.2.

However, in order to understand the solubility phenomena of MWCNTs in different solvents, here we illustrated some possible routes of the solubility mechanism of MWCNTs in water and acetone solvents as shown in Fig. 4.19.

Pristine CNTs could not be soluble in water because of hydrophobic graphite sheet [27] (Fig. 4.16 (Panel I (a))). HNO_3/H_2O_2 O-MWCNTs were insoluble in water (Fig. 4.16 (Panel I (b))), because of their lower contents of surface –COOH groups. Rehman et al. [47] observed a good colloidal HNO_3/H_2O_2 O-MWCNT solution that was lasted for more than one month [47]. The study was not due

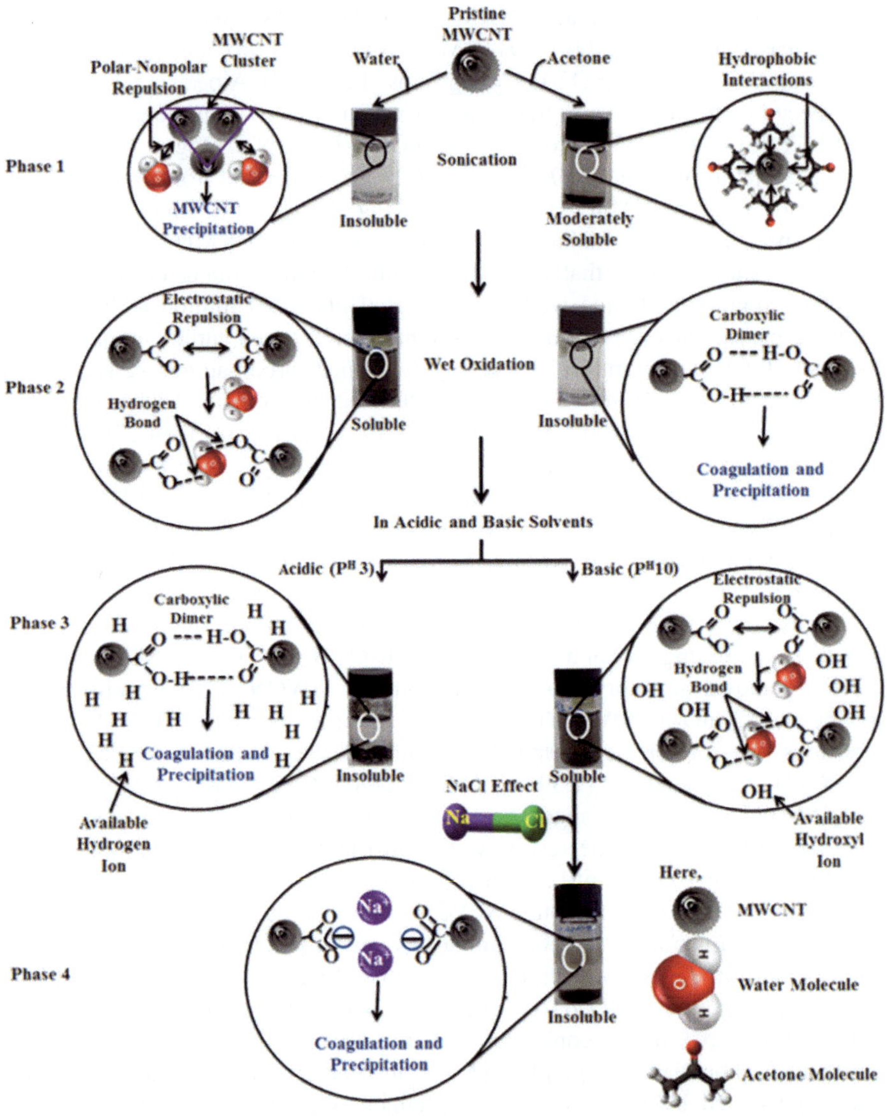

Fig. 4.19 Schematic representation of the proposed dispersibility mechanism of MWCNT in different media. Herein Phase 1, the solubility of pristine MWCNTs in both water and acetone solvents; Phase 2, the behavior of O-MWCNTs in water and acetone solvents; Phase 3, the ionization profiles in acidic pH 3.0 and basic pH 10.0 water solutions and Phase 4, the effects of NaCl on $KMnO_4$ O-MWCNTs in pH 10.0

scrutiny, since most of the findings especially FTIR was misinterpreted. Furthermore, the group dispersed HNO_3/H_2O_2 O-MWCNTs into acetone and water solvents and hypothesized good colloidal dispersibility. A soluble CNT in a protic solution like water cannot be soluble in an aprotic solvent such as acetone.

However, we observed $KMnO_4$ O-MWCNTs showed good solubility (Fig. 4.16 (Panel I (c))) because of their higher content of –COOH groups present on the surface and tip ends. The results were in agreement with XPS and titration results discussed in Sects. 4.4.2 and 4.4.3, respectively. The mechanism could follow two different routes such as (i) carboxylate anions ($-COO^-$) of $KMnO_4$ O-MWCNT may create electrostatic repulsion among the nanotubes, increasing the availability of water contact with the individual MWCNT (Fig. 4.19 (Phase 2)), and (ii) oxygen atom of ($-COO^-$) attacks proton (H^+) of water molecule through electromagnetic dipole–dipole interactions to form intermolecular hydrogen bonds (Fig. 4.19 (Phase 3)). Thus –COOH groups present on O-MWCNTs directly proportionate to the degree of solubility in water and other hydrophilic solvents [71].

In acidic solution, pristine MWCNT was temporarily soluble (Fig. 4.16 (Panel II (a))) might be due to the nucleophilic addition of some water molecules with the epoxide. HNO_3/H_2O_2 O-MWCNTs were not reactive (Fig. 4.16 (Panel II (b))) because of lower –COOH groups. $KMnO_4$ O-MWCNTs were least soluble in acidic (Fig. 4.16 (Panel II (c))), but more soluble in alkaline pH media (Fig. 4.16 (Panel III (c))). It was observed that the rate of deprotonation of –COOH in water was proportional to the increasing pH values [50]. This is due to the availability of protons (H^+) in surrounding acidic environment, that inhibited the formation of ($-COO^-$) and promoted dimerization of –COOH groups which forced O-MWCNTs to agglomerate and precipitate in water solvent (Fig. 4.19 (Phase 3)). In alkaline medium, available –OH groups stimulated –COOH to deprotonate rapidly in order to equilibrate with water pH. Simultaneously, it produces more ($-COO^-$) to interact with the surrounding water molecules.

However, pristine MWCNTs were moderately soluble in acetone solvent (Fig. 4.16 (Panel IV (a))). Herein hydrophobic methyl groups of acetone can interact with hydrophobic graphite sheet of pristine MWCNTs that leads to form a liposome or micelle like structure (Fig. 4.19 (Phase 1)). But the assemblage was not stable because of the presence of highly electronegative surrounding oxygen atoms, which might have tendency to lose liposome or micelle assemblage and subsequently precipitated the pristine MWCNTs as a function of time (Fig. 4.16 (Panel IV (a))). The O-MWCNTs were insoluble (Fig. 4.16 (Panel IV (b and c))) because of the formation of –COOH dimer in contact of the acetone [34, 40]. Herein –COOH groups of O-MWCNTs were less ionized to form intermolecular hydrogen bonds between –COOH groups of O-MWCNTs, but dimerize to precipitate (Fig. 4.19 (Phase 2)).

The contribution of –COOH groups in MWCNT solubility was further justified by adding NaCl solution (Fig. 4.18). Pristine MWCNTs with less –COOH groups were unable to make any electrostatic reaction with salts and precipitated in solution (Fig. 4.18 (Panel I)). $KMnO_4$ O-MWCNTs coagulated and precipitated at the bottom of the vial rapidly (Fig. 4.18 (Panel III)) compared with HNO_3/H_2O_2 O-MWCNTs (Fig. 4.18 (Panel II)) because of the cationic effects of Na^{2+} on –COOH group dissociation. It clearly demonstrated that the higher number of –COOH groups which was present on $KMnO_4$ O-MWCNTs contributed to higher solubility in water solvent. The higher number of –COOH groups rapidly made electrostatic bridges with the higher concentration of Na^{2+} ions. Detail mechanism of this phenomenon is shown in Fig. 4.19 (Phase 4). However, the effect was not reversed by adding acid solution, since ionic interaction depends on the ionic strength of the solution rather than alkaline condition.

4.6 Conclusions

Hydrophobic CNTs have shown to be aggregated and precipitated in polar solvents. These have made their handling difficult and limited their applications in various fields including water purification technologies, catalysis, polymers, composites, sensors, and optoelectronics. Here, we reported two covalent functionalization schemes for MWCNTs using HNO_3/H_2O_2 mixture and basic $KMnO_4$ solution. HNO_3/H_2O_2 O-MWCNTs postulated more oxygen percentage with a high number of C=O and –OH groups, but showed low water solubility. In contrast, $KMnO_4$ O-MWCNTs showed least oxygen percentage, but more –COOH groups that solubilized MWCNTs in water for more than one month. This indicates small changes of surface oxygen concentrations which were depended on a particular oxidizing agent can provoke to generate appropriate functionalities necessary for the dispersibility of O-MWCNTs in water. However, the phenomenon was depended on the solvent pH. In neutral water pH (7.1), –COOH could spontaneously form ($-COO^-$) that equilibrates with surrounding water molecules forming H-bonds and makes homogenous O-MWCNT suspensions. But when the water pH became acidic (3.0), the availability of protons (H^+) in aqueous media inhibited the formation of ($-COO^-$) and promoted dimerization of –COOH groups, which might force O-MWCNTs to agglomerate and precipitate from water solvent. In contrast, available –OH groups in alkaline media (pH 10.0) could stimulate –COOH to deprotonate rapidly in order to equilibrate with water pH, that leads to form more ($-COO^-$) for interacting with the surrounding water molecules. Pristine MWCNTs were insoluble in water because of their hydrophobic graphite sheets, but moderately soluble in acetone solvent. O-MWCNTs were insoluble in acetone solvent because of –COOH dimer formation. Thus, the nature of oxidants and ionic strengths of the solution could play crucial role for manipulating MWCNT solubility in different solvents.

References and Future Readings

1. Abdul Majid, Z., Sabri, M., Anis, N., Buang, N.A., Shahir, S.: Role of oxidant in surface modification of carbon nanotubes for tyrosinase immobilization. J. Funda. Sci. **6**(1), 51–56 (2010)
2. Ahmed, M.H., Byrne, J.A., McLaughlin, J., Elhissi, A., Ahmed, W.: Comparison between FTIR and XPS characterization of amino acid glycine adsorption onto diamond-like carbon (DLC) and silicon doped DLC. Appl. Surf. Sci. **273**, 507–514 (2013)

3. Aitchison, T.J., Ginic-Markovic, M., Matisons, J.G., Simon, G.P., Fredericks, P.M.: Purification, cutting, and sidewall functionalization of multiwalled carbon nanotubes using potassium permanganate solutions. The Journal of Physical Chemistry C **111**(6), 2440–2446 (2007)
4. Andrade, N.F., Martinez, D.S.T., Paula, A.J., Silveira, J.V., Alves, O.L., Souza Filho, A.G.: Temperature effects on the nitric acid oxidation of industrial grade multiwalled carbon nanotubes. J. Nanopart. Res. **15**(7), 1–11 (2013)
5. Baik, S., Usrey, M., Rotkina, L., Strano, M.: Using the selective functionalization of metallic single-walled carbon nanotubes to control dielectrophoretic mobility. J. Phys. Chem. B **108** (40), 15560–15564 (2004)
6. Biniak, S., Szymański, G., Siedlewski, J., Światkowski, A.: The characterization of activated carbons with oxygen and nitrogen surface groups. Carbon **35**(12), 1799–1810 (1997)
7. Bokobza, L., Zhang, J.: Raman spectroscopic characterization of multiwall carbon nanotubes and of composites. Express Polym. Lett **6**, 601–608 (2012)
8. Bortolamiol, T., Lukanov, P., Galibert, A-M., Soula, B., Lonchambon, P., Datas, L., Flahaut, E.: Double-walled carbon nanotubes: quantitative purification assessment, balance between purification and degradation and solution filling as an evidence of opening. Carbon **78**, 79–90 (2014)
9. Chen, G.-X., Kim, H.-S., Park, B.H., Yoon, J.-S.: Controlled functionalization of multiwalled carbon nanotubes with various molecular-weight poly (l-lactic acid). J. Phys. Chem. B **109** (47), 22237–22243 (2005)
10. Chen, J., Hamon, M.A., Hu, H., Chen, Y., Rao, A.M., Eklund, P.C., Haddon, R.C.: Solution properties of single-walled carbon nanotubes. Science **282**(5386), 95–98 (1998)
11. Das, R., Ali, M.E., Hamid, S.B.A., Ramakrishna, S., Chowdhury, Z.Z.: Carbon nanotube membranes for water purification: a bright future in water desalination. Desalination **336**, 97–109 (2014)
12. Das, R., Hamid, S.B.A., Ali, M.E., Ismail, A.F., Annuar, M., Ramakrishna, S.: Multifunctional carbon nanotubes in water treatment: the present, past and future. Desalination **354**, 160–179 (2014)
13. Datsyuk, V., Kalyva, M., Papagelis, K., Parthenios, J., Tasis, D., Siokou, A., Kallitsis, I., Galiotis, C.: Chemical oxidation of multiwalled carbon nanotubes. Carbon **46**(6), 833–840 (2008)
14. Dillon, A.C., Gennett, T., Jones, K.M., Alleman, J.L., Parilla, P.A., Heben, M.J.: A simple and complete purification of single-walled carbon nanotube materials. Adv. Mater. **11**(16), 1354–1358 (1999)
15. Eckmann, A., Felten, A., Mishchenko, A., Britnell, L., Krupke, R., Novoselov, K.S., Casiraghi, C.: Probing the nature of defects in graphene by Raman spectroscopy. Nano Lett. **12**(8), 3925–3930 (2012)
16. Eklund, P., Holden, J., Jishi, R.: Vibrational modes of carbon nanotubes; spectroscopy and theory. Carbon **33**(7), 959–972 (1995)
17. Esteve, W., Budzinski, H., Villenave, E.: Relative rate constants for the heterogeneous reactions of OH, NO_2 and NO radicals with polycyclic aromatic hydrocarbons adsorbed on carbonaceous particles. Part 1: PAHs adsorbed on 1–2 μm calibrated graphite particles. Atmos. Environ. **38**(35), 6063–6072 (2004)
18. Feng, W., Ji, P.: Enzymes immobilized on carbon nanotubes. Biotechnology advances **29**(6), 889–895 (2011)
19. Fogden, S., Verdejo, R., Cottam, B., Shaffer, M.: Purification of single walled carbon nanotubes: the problem with oxidation debris. Chem. Phys. Lett. **460**(1), 162–167 (2008)
20. Gao, Y., Kyratzis, I.: Covalent immobilization of proteins on carbon nanotubes using the cross-linker 1-ethyl-3-(3-dimethylaminopropyl) carbodiimide—a critical assessment. Bioconjugate Chem **19**(10), 1945–1950 (2008)
21. Grandi, S., Magistris, A., Mustarelli, P., Quartarone, E., Tomasi, C., Meda, L.: Synthesis and characterization of SiO_2–PEG hybrid materials. J. Non-Cryst. Solids **352**(3), 273–280 (2006). doi:10.1016/j.jnoncrysol.2005.11.033

22. Gupta, S.S., Sreeprasad, T.S., Maliyekkal, S.M., Das, S.K., Pradeep, T.: Graphene from sugar and its application in water purification. ACS Appl. Mater. Interfaces. **4**(8), 4156–4163 (2012)
23. Hernadi, K., Siska, A., Thien-Nga, L., Forro, L., Kiricsi, I.: Reactivity of different kinds of carbon during oxidative purification of catalytically prepared carbon nanotubes. Solid State Ionics **141**, 203–209 (2001)
24. Hill, D.E., Lin, Y., Rao, A.M., Allard, L.F., Sun, Y.-P.: Functionalization of carbon nanotubes with polystyrene. Macromolecules **35**(25), 9466–9471 (2002)
25. Hiura, H., Ebbesen, T.W., Tanigaki, K.: Opening and purification of carbon nanotubes in high yields. Adv. Mater. **7**(3), 275–276 (1995)
26. Hontoria-Lucas, C., Lopez-Peinado, A., López-González, JdD, Rojas-Cervantes, M., Martin-Aranda, R.: Study of oxygen-containing groups in a series of graphite oxides: physical and chemical characterization. Carbon **33**(11), 1585–1592 (1995)
27. Hou, P.-X., Liu, C., Cheng, H.-M.: Purification of carbon nanotubes. Carbon **46**(15), 2003–2025 (2008)
28. Huang, T., Tzeng, Y., Liu, Y., Chen, Y., Walker, K., Guntupalli, R., Liu, C.: Immobilization of antibodies and bacterial binding on nanodiamond and carbon nanotubes for biosensor applications. Diam. Relat Mater. **13**(4), 1098–1102 (2004)
29. Huang, W., Taylor, S., Fu, K., Lin, Y., Zhang, D., Hanks, T.W., Rao, A.M., Sun, Y.-P.: Attaching proteins to carbon nanotubes via diimide-activated amidation. Nano Lett. **2**(4), 311–314 (2002)
30. Huang, Y.-L., Tien, H.-W., Ma, C.-C.M., Yang, S.-Y., Wu, S.-Y., Liu, H.-Y., Mai, Y.-W.: Effect of extended polymer chains on properties of transparent graphene nanosheets conductive film. J. Mater. Chem. **21**(45), 18236–18241 (2011)
31. Kenkel, J.: Analytical chemistry for technicians. CRC Press, Boca Raton (2002)
32. Kim, U.J., Furtado, C.A., Liu, X., Chen, G., Eklund, P.C.: Raman and IR spectroscopy of chemically processed single-walled carbon nanotubes. J. Am. Chem. Soc. **127**(44), 15437–15445 (2005)
33. Klyushin, A.Y., Rocha, T.C., Hävecker, M., Knop-Gericke, A., Schlögl, R.: A near ambient pressure XPS study of Au oxidation. PCCP **16**(17), 7881–7886 (2014)
34. Li, M., Boggs, M., Beebe, T.P., Huang, C.: Oxidation of single-walled carbon nanotubes in dilute aqueous solutions by ozone as affected by ultrasound. Carbon **46**(3), 466–475 (2008)
35. Lourderaj, U., Giri, K., Sathyamurthy, N.: Ground and excited states of the monomer and dimer of certain carboxylic acids. J. Phys. Chem. A **110**(8), 2709–2717 (2006)
36. Matarredona, O., Rhoads, H., Li, Z., Harwell, J.H., Balzano, L., Resasco, D.E.: Dispersion of single-walled carbon nanotubes in aqueous solutions of the anionic surfactant NaDDBS. J. Phys. Chem. B **107**(48), 13357–13367 (2003)
37. Mattevi, C., Eda, G., Agnoli, S., Miller, S., Mkhoyan, K.A., Celik, O., Mastrogiovanni, D., Granozzi, G., Garfunkel, E., Chhowalla, M.: Evolution of electrical, chemical, and structural properties of transparent and conducting chemically derived graphene thin films. Adv. Funct. Mater. **19**(16), 2577–2583 (2009)
38. Mishra, G., McArthur, S.L.: Plasma polymerization of maleic anhydride: just what are the right deposition conditions? Langmuir **26**(12), 9645–9658 (2010)
39. Moulder, J., Stickle, W., Sobol, P., Bomben, K.: Handbook of X-ray photoelectron spectroscopy; Perkin-Elmer Corporation: Waltham, MA (1992)
40. Nandi, C.K., Hazra, M.K., Chakraborty, T.: Vibrational coupling in carboxylic acid dimers. J. Chem. Phys. **123**(12), 124310 (2005)
41. O'connell, M.J., Bachilo, S.M., Huffman, C.B., Moore, V.C., Strano, M.S., Haroz, E.H., Rialon, K.L., Boul, P.J., Noon, W.H., Kittrell, C.: Band gap fluorescence from individual single-walled carbon nanotubes. Science **297**(5581), 593–596 (2002)
42. Osswald, S., Havel, M., Gogotsi, Y.: Monitoring oxidation of multiwalled carbon nanotubes by Raman spectroscopy. J. Raman Spectrosc. **38**(6), 728–736 (2007)
43. Parveen, S., Rana, S., Fangueiro, R.: A review on nanomaterial dispersion, microstructure, and mechanical properties of carbon nanotube and nanofiber reinforced cementitious composites. J. Nanomaterials **2013**, 80 (2013)

44. Pelalak, R., Baniadam, M., Maghrebi, M.: Controllable purification, cutting and unzipping of multi-walled carbon nanotubes with a microwave method. Appl. Phys. A **111**(3), 951–957 (2013)
45. Peng, Y., Liu, H.: Effects of oxidation by hydrogen peroxide on the structures of multiwalled carbon nanotubes. Ind. Eng. Chem. Res. **45**(19), 6483–6488 (2006)
46. Qin, Y., Liu, L., Shi, J., Wu, W., Zhang, J., Guo, Z.-X., Li, Y., Zhu, D.: Large-scale preparation of solubilized carbon nanotubes. Chem. Mater. **15**(17), 3256–3260 (2003)
47. Rehman, A.U., Abbas, S.M., Ammad, H.M., Badshah, A., Ali, Z., Anjum, D.H.: A facile and novel approach towards carboxylic acid functionalization of multiwalled carbon nanotubes and efficient water dispersion. Mater. Lett. **108**, 253–256 (2013)
48. Rück-Braun, K., Petersen, M.Å., Michalik, F., Hebert, A., Przyrembel, D., Weber, C., Ahmed, S.A., Kowarik, S., Weinelt, M.: Formation of carboxy-and amide-terminated alkyl monolayers on silicon (111) investigated by ATR-FTIR, XPS, and X-ray scattering: construction of photoswitchable surfaces. Langmuir **29**(37), 11758–11769 (2013)
49. Q-j, Shen, X-b, Liu, W-j, Jin: Solubility increase of multi-walled carbon nanotubes in water. Carbon **60**, 562–563 (2013)
50. Shieh, Y.-T., Liu, G.-L., Wu, H.-H., Lee, C.-C.: Effects of polarity and pH on the solubility of acid-treated carbon nanotubes in different media. Carbon **45**(9), 1880–1890 (2007)
51. Shin, H.J., Kim, K.K., Benayad, A., Yoon, S.M., Park, H.K., Jung, I.S., Jin, M.H., Jeong, H. K., Kim, J.M., Choi, J.Y.: Efficient reduction of graphite oxide by sodium borohydride and its effect on electrical conductance. Adv. Funct. Mater. **19**(12), 1987–1992 (2009)
52. Smith, B., Wepasnick, K., Schrote, K.E., Cho, H.-H., Ball, W.P., Fairbrother, D.H.: Influence of surface oxides on the colloidal stability of multi-walled carbon nanotubes: a structure—property relationship. Langmuir **25**(17), 9767–9776 (2009)
53. Spencer, N.D.: Tailoring surfaces: modifying surface composition and structure for applications in tribology, biology and catalysis, vol. 5. World scientific, Singapore (2011)
54. Subrizi, F., Crucianelli, M., Grossi, V., Passacantando, M., Pesci, L., Saladino, R.: Carbon nanotubes as activating tyrosinase supports for the selective synthesis of catechols. ACS Catal. **4**(3), 810–822 (2014)
55. Sun, Y.-P., Fu, K., Lin, Y., Huang, W.: Functionalized carbon nanotubes: properties and applications. Acc. Chem. Res. **35**(12), 1096–1104 (2002)
56. Tang, M., Dou, H., Sun, K.: One-step synthesis of dextran-based stable nanoparticles assisted by self-assembly. Polymer **47**(2), 728–734 (2006). doi:10.1016/j.polymer.2005.11.091
57. Tasis, D., Tagmatarchis, N., Bianco, A., Prato, M.: Chemistry of carbon nanotubes. Chemical reviews **106**(3), 1105–1136 (2006)
58. Thomas, S., Stephen, R.: Rubber Nanocomposites. Preparation, Properties, and Applications, vol. 10. John Wiley & Sons Frontiers in Materials and Minerals Engineering, Singapore (2010)
59. Tian, R., Wang, X., Li, M., Hu, H., Chen, R., Liu, F., Zheng, H., Wan, L.: An efficient route to functionalize singe-walled carbon nanotubes using alcohols. Appl. Surf. Sci. **255**(5), 3294–3299 (2008)
60. Vickerman, J.C., Gilmore, I.S.: Surface analysis: the principal techniques, vol. 2. Wiley Online Library (2009)
61. Vione, D., Maurino, V., Minero, C., Pelizzetti, E., Harrison, M.A., Olariu, R.-I., Arsene, C.: Photochemical reactions in the tropospheric aqueous phase and on particulate matter. Chem. Soc. Rev. **35**(5), 441–453 (2006)
62. Walling, C.: Fenton's reagent revisited. Acc. Chem. Res. **8**(4), 125–131 (1975). doi:10.1021/ar50088a003
63. Wang, J., Zhao, G., Li, Y., Zhu, H., Peng, X., Gao, X.: One-step fabrication of functionalized magnetic adsorbents with large surface area and their adsorption for dye and heavy metal ions. Dalton Trans. **43**(30), 11637–11645 (2014)
64. Wang, Y., Shan, H., Hauge, R.H., Pasquali, M., Smalley, R.E.: A highly selective, one-pot purification method for single-walled carbon nanotubes. J. Phys. Chem. B **111**(6), 1249–1252 (2007)

65. Weber, W.H., Merlin, R. (2000). Raman scattering in materials science, vol. 42. Springer, Berlin
66. Wepasnick, K.A., Smith, B.A., Schrote, K.E., Wilson, H.K., Diegelmann, S.R., Fairbrother, D.H.: Surface and structural characterization of multi-walled carbon nanotubes following different oxidative treatments. Carbon **49**(1), 24–36 (2011)
67. Yang, N., Zoski, C.G.: Polymer films on electrodes: investigation of ion transport at poly (3,4-ethylenedioxythiophene) films by scanning electrochemical microscopy. Langmuir **22** (25), 10338–10347 (2006)
68. Zeng, L., Zhang, L., Barron, A.R.: Tailoring aqueous solubility of functionalized single-wall carbon nanotubes over a wide pH range through substituent chain length. Nano Lett. **5**(10), 2001–2004 (2005)
69. Zhang, J., Zou, H., Qing, Q., Yang, Y., Li, Q., Liu, Z., Guo, X., Du, Z.: Effect of chemical oxidation on the structure of single-walled carbon nanotubes. J. Phys. Chem. B **107**(16), 3712–3718 (2003)
70. Zhang, Z., Zhang, T., Li, J., Ji, Z., Zhou, H., Zhou, X., Gu, N.: Preparation of poly (l-lactic acid)-modified polypropylene mesh and its antiadhesion in experimental abdominal wall defect repair. J. Biomed. Mater. Res. B Appl. Biomater. **102**(1), 12–21 (2014)
71. Zhao, W., Song, C., Pehrsson, P.E.: Water-soluble and optically pH-sensitive single-walled carbon nanotubes from surface modification. J. Am. Chem. Soc. **124**(42), 12418–12419 (2002)

Chapter 5
Nanobiohybrid Preparation

Abstract Enzyme immobilization onto nanomaterials has been implemented in various fields such as water decontamination, sensor developments, biotransformation, therapeutics, foods processing, biofuel production, and so on. In this study, we aimed to covalently immobilize 3,4-POD onto H_2SO_4 and HNO_3 functionalized (F)-MWCNTs to birth Nanobiohybrid catalyst. Images of SEM, TEM, and AFM along with UV/vis and IR spectroscopic data demonstrated that the 3,4-POD was successfully immobilized onto F-MWCNT surfaces. CD spectroscopy data showed that the Nanobiohybrid undergone 44% of relative structural changes to its free 3,4-POD configurations. Optimizing immobilization parameters, such as the use of cross-linker, time incubation, and different concentrations of 3,4-POD loading helped us to attach maximum 1060 μg of 3,4-POD/mg of MWCNT. This paves the way for the development of effective Nanobiohybrid that might have the imminent potentiality to purify 3,4-DHBA contaminated wastewater.

5.1 Background

Potential microbial decontamination (MD) of wastewater has been historically popular and studied in great details [27, 69]. Of particular interest is the biotransformation of notorious water pollutants that has been actually carried out by enzymes inside the microbial cells. It has been applied to treat complex wastewater matrices, where chemical treatments are typically inoperative. However, MD has been facing some shortcomings such as high costs, time-consuming microbial growth processes, extreme sensitivity toward pH, temperatures, and stress, predators, toxic byproduct formations, dependence on nutrient growth mediums, and generation of biomass which make MD dormant to handle both the conventional priorities and newly emerging water pollutants [39].

R. Das, *Nanohybrid Catalyst based on Carbon Nanotube*,
Carbon Nanostructures, DOI 10.1007/978-3-319-58151-4_5

In an attempt to prevail these shortcomings, water scientists have been budged to separate enzyme catalysts from their parental microorganisms and attenuated different wastewater pollutants [28, 46]. It has been popular because of its advantages such as mild reaction conditions, high effectiveness and specificity, high owing to low operating costs, and time, absence of secondary reactions, low or no energy consumption, and finally eco-friendly than physical, chemical, and biological counterparts [26, 39, 62]. But free enzyme has evinced poor stability and lifetime, high sensitivity to mechanical stresses, and difficult to separate from its substrates and products [7, 22, 26]. With the purposes of increasing their stability and recrudescence, immobilization of enzymes onto physical supports has been well documented [26, 62, 75].

The enzyme 3,4-POD (EC 1.13.1.3) is an intradiol cleaving metalloenzyme that consists of nonheme Fe (III) at the active site. It catalyzes the transformation of an emerging toxic water pollutants 3,4-DHBA [2, 61] to nontoxic 3-carboxy-cis, cis-muconic acid [47, 49] as shown in Fig. 5.1. The enzyme has been widely purified from *P. putida* [24] and became a popular industrial catalyst because of its direct oxidation of 3,4-DHBA without any adjuvant. Immobilizing 3,4-POD onto MWCNT would have certain benefits. First, the hydrogenation of muconic acid could persuade adipic acid that has commercially been used as a monomer for producing nylon, elastomers, flavorant or gelling aid in foods or pharmaceuticals, cloths, polyurethane, and tissue engineering as shown in Fig. 5.1 [19, 31, 40]. These have obligated to produce 2.3 million metric tons adipic acid worldwide in 2012, which had market value of 6.3 billion pounds [19, 57]. Second, chemical synthesis of 3-carboxy-cis,cis-muconic acid has been difficult [73, 74], which could also be commercially utilized for producing common ternary chemicals (Fig. 5.1) [40]. Finally, attaching 3,4-POD onto F-MWCNT surface would facilitate the dispersibility of Nanobiohybrid, since the enzyme contains various polar residues on its surface to react with water molecules. Matsuura et al. [44] proposed an enzyme immobilized on CNT solubility could increase Nanobiohybrid solubility by exposing enzyme's hydrophilic domain on its surface [44]. With these overall

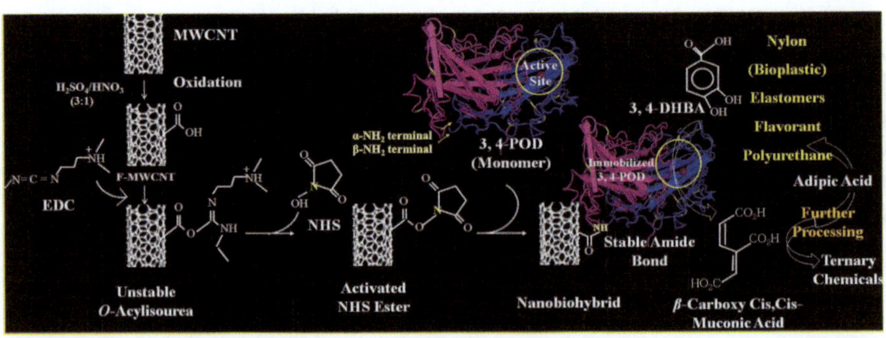

Fig. 5.1 Synthesis of Nanobiohybrid catalyst as depicted in a schematic representation of the conjugation of 3,4-POD to F-MWCNTs using EDC

pictures in mind, we commenced to immobilize 3,4-POD onto a support such as MWCNTs for improving its stability and efficiency compared with free enzyme.

A few studies have been attempted to immobilize 3,4-POD on porous glass [73, 74], agarose [28, 76], and alginate matrix [28, 58]. However, the studies have found without due scrutiny, since crude 3,4-POD extracts have been used [28, 58]. Moreover, no characterizations of immobilized structures were done; so the data should be used with cautions to understand its observed catalytic efficiency. In addition, immobilized 3,4-POD has not sufficiently been tested kinetically [28, 58, 73, 74, 76], making it difficult for the scientists to understand accurate 3,4-POD behaviors at support interfaces.

However, to date, no study has yet been published or attempted to immobilize pure 3,4-POD on nanomaterials. The judicious choices that have been influenced us to select MWCNTs as support because of its (i) high surface area, (ii) large aspect ratio, (iii) ease CNT fabrication and operational advantages, (iv) low preparation costs, and (v) wide availability [10, 22]. In addition, CNTs have shown to be stable and inert for microbial contamination that might give immobilized 3,4-POD prolong activity, antifouling, self-cleaning, and reusable functions in wastewater treatment plant [17]. Alternatively, a hollow CNT could provide frictionless transport of water molecules, and this makes them suitable for the development of high fluxing separation techniques and would provide low energy solution for water treatment [13]. Moreover, it has been shown that the CNTs could be a better carrier of enzyme immobilization than glassy carbons in terms of structure and conformational changes of attached enzymes [32].

In this study, we immobilized 3,4-POD onto MWCNTs using the cross-linker 1-ethyl-3-(3-dimethylaminopropyl)carbodiimide (EDC) reactions as shown in Fig. 5.1.

However, hydrophobic CNTs have shown to be aggregated and precipitated in water, which make their handling difficult and limit their applications in enzyme immobilizations [41, 62] and water purifications [15]. In order to get a suitable MWCNT matrix, at first we compared the colloidal stabilities of $KMnO_4$ O-MWCNTs [14] with H_2SO_4 and HNO_3 functionalized (F)-MWCNTs [5]. Finally, we successfully developed a Nanobiohybrid catalyst by optimizing three immobilizing reaction conditions such as the use of EDC, time incubations, and variable quantities of 3,4-POD loading onto a highly soluble F-MWCNT surface.

5.2 Literature Review

Nanobiohybrid has been routinely utilized in water purification [75], sensing [42], therapeutics [54], foods processing [77], biodiesels [66], petroleum refining [70], and so on. It has been made by immobilizing enzymes onto various supports such as CNTs, nanoparticles, nanofibers, and so on via non-covalent and covalent immobilization techniques [33, 65]. Covalent enzyme conjugation has been popular because of its strong interactions with the support, which allows a durable enzyme attachment during operation compared with non-covalent enzyme conjugation [22]. It has been hypothesized that the covalent bond could tune the locomotion of enzyme molecules which is important to give high stability, selectivity, accessibility, and reduce leaching [36]. However, covalent immobilization could potentially change the enzyme and/or protein structures upon immobilization compared with non-covalent conjugation processes [22]. It has been argued that no matter what immobilization techniques have been chosen as long as one could maintain stable attachment and improve stability and functions to their innate state [52].

5.2.1 Enzyme Immobilization onto CNT Using Non-covalent Approach

Non-covalent methods of enzyme immobilization typically preserve the integrity of CNT and also the conformation of enzyme molecules. Enzymes typically adsorb on the surface of CNT in a localized or global scale. Adsorption is a phenomenon where enzymes can be physically adsorbed into the CNTs. Herein, enzymes and CNT are soaked in a bath and allow time for the physical interaction between them. After that, the unbound enzymes can be washed out using buffer solution. Various polymers and linking molecules can be used to hold CNT at one end and enzymes at other ends.

5.2.1.1 Physical Adsorption

Similar to non-covalent CNT functionalization approach that we discussed in previous Chap. 4, enzyme can be directly adsorbed into the CNT using hydrophobic, π–π, electrostatic, and hydrogen bonding interactions. First, those enzymes whose have exterior hydrophobic amino acid residues can be directly sorbed into the hydrophobic surface of CNT. Second, π–π stacking between the aromatic residues of enzymes and CNTs contribute for spontaneous adsorption. Third, some enzymes, e.g., lysozyme can interact with the CNT using electrostatic bond, where amino groups of enzymes should be protonated first (i.e., pH < isoelectric point) which will create electrostatic bond with the defected sites of CNT. At last, since most of the functionalized CNT have COOH group, the amino acid residues in enzymes can create hydrogen bond with COOH. These bonding actions and there nature are very important to determine the amount of enzymes immobilized onto CNT and maximize their activity at nano–bio interface. It should be kept in mind that most of the scientists have used sonicator for enzyme adsorption into CNT. Sonication should be avoided, since it creates a large pressure gradient that could denature the enzyme molecules reduce its activity.

5.2.1.2 Polymers Mediated Enzyme Immobilization

As we discussed in preceding Chap. 4, the polymers and biomolecules functionalized CNT might have good water dispersibility, and thence the immobilization of enzyme onto CNT can be facilitated. Colloidal stability of CNT is an important issue to bring the enzyme molecule on its surface for attachment. Both the polymers and biomolecules coated CNT yield positively and negatively charged CNTs which can interact with the enzyme molecules through electrostatic bonds. For example, CNT can be positively charged by functionalizing poly(sodium 4-styrenesulfonate) combined with ionic liquids. Chitosan is another popular polymer which yields positively charged CNT.

5.2.1.3 Biomolecules Mediated Enzyme Immobilization

Biomolecules such as DNA functionalized CNT can be used for enzyme immobilization. First of all, chemical attaching of single-stranded DNA (ssDNA) with CNT should be done for hybridization with complementary DNA (cDNA) chains. The anchoring of DNA onto CNT primarily occurred by energetically favorable π–π stacking interaction, where the plane of the aromatic nucleotide bases oriented parallel to the surface of the nanotube. After wrapping of ssDNA on CNTs, enzymes can be immobilized onto DNA-wrapped CNTs. The most important role for the DNA link is enabling the site-addressable binding of redox proteins to CNT. Enzymes labeled with the cDNA tags can recognize their binding address and adhere to the CNTs functionalized with the complementary ssDNA strands, as illustrated by Fig. 5.2. Sometimes some linking molecules such as streptavidin,

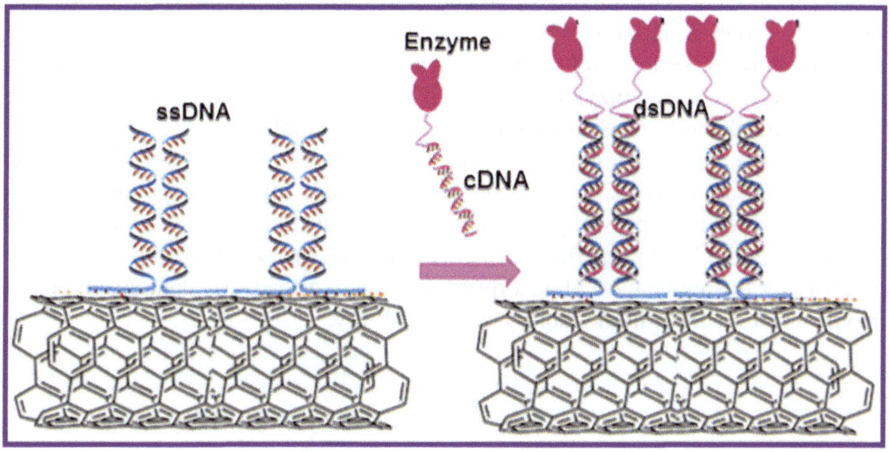

Fig. 5.2 Conjugation of enzymes tagged with single-stranded DNA to the tips of designated CNTs functionalized with the complementary strands

flavin adenine dinucleotide cofactor can be used between enzymes and CNTs. Since enzyme will be at distal phase from the CNT near surface, the conformation of enzyme molecule will be preserved as compared to direct enzyme attachment on CNT surface. Therefore, DNA-mediated enzyme attachment can yield high enzyme activity, and moreover, molecular recognition and accessibility of substrate into such immobilized enzymes could also be accelerated.

5.2.1.4 Surfactant Mediated Enzyme Immobilization

Surfactant often used to solubilize the CNT in aqueous solution. Surfactant (e.g., Triton X-100) mediated enzyme immobilization is specific and has been shown as efficient method. Enzymes, such as horseradish peroxidase and cytochrome *c*, can be easily co-assembled with sodium dodecyl sulfate, cetyltrimethylammonium bromide, and Triton X-100 surfactants onto CNT. Co-adsorption of Triton X-100 with poly(ethylene glycol) onto CNTs can be used for selective immobilization of enzymes. For the adsorption of enzyme, CNTs are first sonicated in the solution of a surfactant for better dispersibility. The suspension is then mixed with enzyme, which facilitates enzyme adsorption more strongly due to well-dispersed CNT solution where enzymes could gain well access to the individual CNT molecule.

5.2.1.5 Layer-by-Layer Approach for Enzyme Immobilization

Layer-by-layer (LBL) approach is an effective method for multiple enzymes immobilization. Producing multilayer enzymes film onto CNT is often necessary for multifunctional catalyst development. Moreover, the catalytic activity of enzyme can be increased by increasing the enzyme concentration assembled into the layers on CNT. One mighty approach is to functionalize the CNT using acids for COO^- group generation. Then the positively charged poly(dimethyldiallylammoniumchloride), simply called PDDA is assembled, followed by the assembling of negatively charged enzyme. Figure 5.3 reveals a stepwise LBL assembly of alkaline phosphatase (ALP) multilayers on a CNT template.

The above approaches of enzyme immobilization onto the CNT using non-covalent approaches, i.e., adsorption methods, might carry some unique advantages, such as improved preservation of enzyme structures, intrinsic electronic structures, and CNT properties. However, durability and leaching are always concerns for the application of adsorption methods. Instead, covalent attachment of enzymes can result in high stability and reusability.

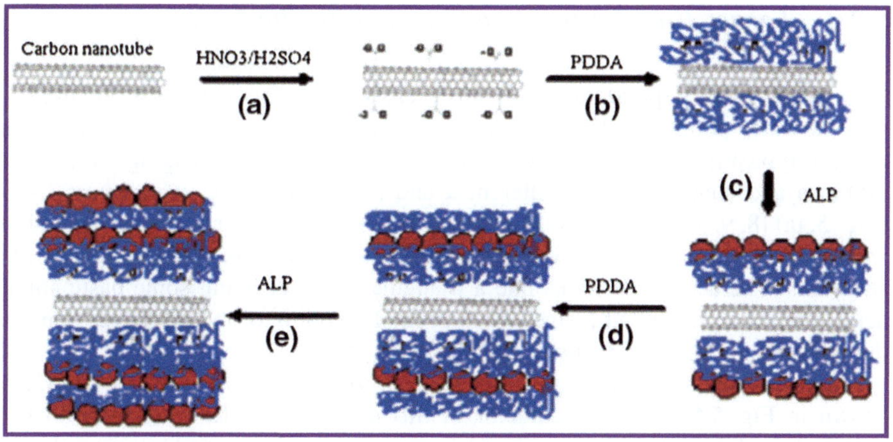

Fig. 5.3 Graphical representation of LBL approach where electrostatic self-assembly of alkaline phosphatase polyion on CNT template. *PDDA* poly (diallyldimethylammonium chloride) polymer; *ALP* alkaline phosphatase. Adapted with permission from ACS [45]

5.2.2 Enzyme Immobilization onto CNT Using Covalent Approach

We already discussed the detail method of covalent enzyme immobilization onto CNT in the Sect. 5.1. In short, the method involves the reaction between the free NH_2 group of enzyme and COOH groups that have been anchored during CNT oxidation. The reaction principle follows carbodiimide which has been commonly used for Nanobiohybrid development as shown in Fig. 5.1. Such system is stable at high temperatures, providing a unique combination of useful attributes such as low mass transfer resistance, high activity and stability, and reusability. Another advantage is that most of the enzymes might immobilized locally onto the CNT which may promote the accessibility of substrate to the enzyme and the CNTs can facilitate heat transfer. It has been demonstrated that enzymes on SWCNTs have much higher activity than those conjugated to MWCNTs [52]. Sometimes linking molecules such as 1-pyrenebutanoic acid succinimidyl ester, the high reactive succinimidy ester groups were covered on CNTs, facilitating the horseradish peroxidase immobilization. Aminopyrene is another linking molecule where the amino functional groups can be introduced uniformly on the CNT surface, and the immobilized laccase showed higher electrocatalytic activity and better stability than the laccase immobilized on the pristine CNTs.

5.2.3 Target Enzyme for Attachment with the CNT

One of the most popular intradiol dioxygenase enzymes that has been used in various water pollutants degradation is 3,4-POD [29] as shown in Fig. 5.4. The enzyme typically exists as heterodimer with molecular mass ranging from 97 to 700 kDa and consists of two different α and β subunits $(\alpha\beta)_n$, where n = (2–12) (Fig. 5.4a) [8, 9, 11, 30, 79]. Both subunits have shown similar folds surrounded by small helices and consist of two four-strand-β-sheets that make them sandwich like structures. The active site looks like a triangular channel lining some basic amino acids such as arginine and lysine which promote binding of the 3,4-DHBA and its derivatives into the active site of 3,4-POD (Fig. 5.4b).

The mechanism of action of 3,4-POD catalyzes the cleavage of 3,4-DHBA as shown in Fig. 5.5 [67]. The cofactor of this enzyme Fe (III) attached with two tyrosines and two histidines amino acids which have played major role for degrading 3,4-DHBA. Therefore, stabilizing its catalytic efficiencies by immobilizing the 3,4-POD onto a support matrix (i.e., Nanobiohybrid) would give a suitable and effective novel way for purifying 3,4-DHBA contaminated wastewater effluents.

Although several substrate specificity and regioselectivity of 3,4-POD have been observed for 1,2-dihydroxybenzene, 3-methylcatechol, 4-methylcatechol, 1,2,3-tri hydroxybenzene [24], 4-sulfocatechol [30], 2,3-dihydroxybenzoate, 2,4-dihydroxy benzoate, caffeic acid, 3,4-dihydroxycinnamic acid, and so on [28]; 100% relative activity of 3,4-POD was only obtained for 3,4-DHBA than other co-substrate mixtures [24]. Although 3,4-POD activity toward other polyphenols was lower, one

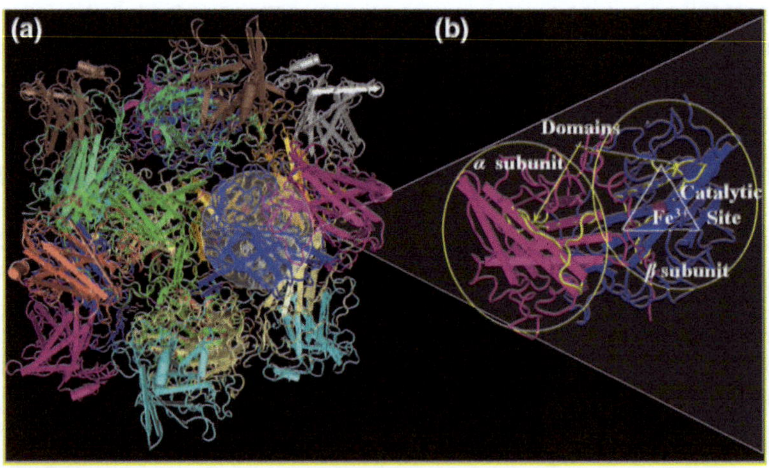

Fig. 5.4 Tertiary 3,4-POD structure **a** symmetry (T) view of $(\alpha\beta)_{12}$ protomers retrieved from the protein data bank (PDB) with following ID, 3PCC; and **b** asymmetry view of monomer $(\alpha\beta)_1$ with PDB ID, 2PCD

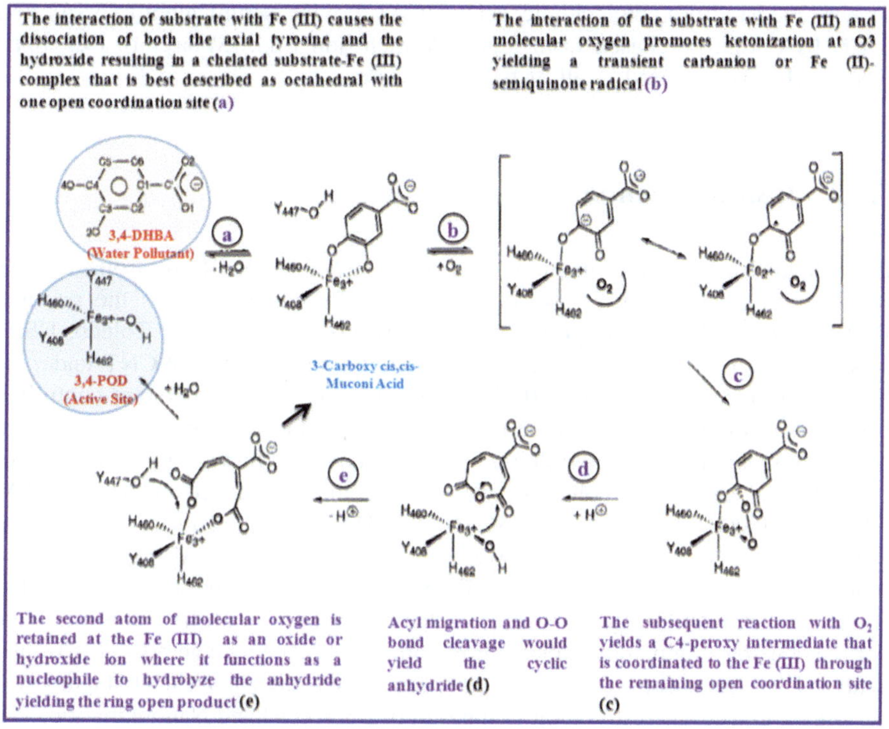

Fig. 5.5 Schematic representation of summary of catalytic degradation of 3,4-DHBA by 3,4-POD active site. Figure is adapted with permission from ACS [67]

can still envision for eliminating them using a single Nanobiohybrid that would accelerate the degradation rate with reduced cost in the near future.

5.3 Materials and Methods

5.3.1 Materials and Reagents

MWCNTs of 12 ± 5 and 4.0 nm in outer and inner diameters and >1 μm in length were purchased from Bayer MaterialScience AG (Germany). The tubes were prepared by catalytic CVD and contained >95% carbon by weight and were used as received. Lyophilized powder of pure 3,4-POD ($\geq$3 units/mg solid) from *Pseudomonas* sp. was purchased from Sigma-Aldrich Sdn Bhd. (Malaysia) and used without further purification. HCl (37%), H_2O_2 (30%), H_2SO_4 (98%), HNO_3 (65%), sodium hydroxide, ethanol (70%), 2-(*N*-morpholino)ethanesulfonic acid (MES), EDC, N-hydroxysuccinimide (NHS), 4-(2-hydroxyethyl)-1-piperazineethanesulfonic acid (HEPES), N-Cyclohexyl-2-aminoethanesulfonic acid (CHES),

Tween 20, Bradford, bicinchonic acid (BCA), bovine serum album (BSA) and 3,4-DHBA were purchased from Sigma-Aldrich Sdn Bhd. (Malaysia). Deionized water (pH 6.8) was used in all of the experiments.

5.3.2 *Instrumentation*

In order to determine the functionalities of MWCNT, ATR-IR spectrometer (IFS 66 v/S, Bruker, Germany) was conducted on KBr embedded into the pristine, F-MWCNTs, and Nanobiohybrid. TGA (TGA/SDTA 851, Mettler Toledo, USA) was performed to check the weight loss of the pristine and F-MWCNT under air flow (50 ml) between 25 and 1000 °C at 10 °C/min. A temperature-controlled UV/vis spectrophotometer (V-630, JASCO, Japan) was used for measuring MWCNT solubility and determining 3,4-POD activities. Surface morphologies of F-MWCNTs and Nanobiohybrid were successfully checked using FE-SEM (Hitachi-SU8000, Japan) and TEM (Hitachi-HT7700 at 120 kV, Japan). The F-MWCNTs and Nanobiohybrid were dispersed into fresh MilliQ water and mounted onto lacey copper grids for FE-SEM and TEM analyses. AFM (Bruker BioScope Catalyst, Germany) images of Nanobiohybrid on a glass substrate were collected using commercial silicon tips with a frequency range 51–94 kHz and analyzed by Nanoscope software. CD spectroscopy (J-810, JASCO, Japan) was performed to determine structural changes of free 3,4-POD and Nanobiohybrid. The sample solutions having 0.05 mg/ml of either free 3,4-POD or Nanobiohybrid in CHES buffer (10 mM, pH 5.6) from which 300 μL was used for CD analyses. F-MWCNT solution (15 μg/ml) that is equivalent with the concentration of Nanobiohybrid support was used as control in the same buffer solution. All readings were collected from three consecutive scans for each CD spectra.

5.3.3 *Preparation of F-MWCNT*

About 0.5 g of HCl/H_2O_2 purified MWCNTs [16] was further functionalized with 8.0 ml mixture of H_2SO_4 and HNO_3 (3:1 v/v) [5]. The mixture was sonicated at 50 °C for 8 h in an ultrasonication bath (Series 400; Powersonic; 40 kHz; Korea). The F-MWCNTs were extracted from the residual acids, bases, metallic byproducts, and carbonaceous impurities by repeated cycle of dilutions with fresh double-distilled water followed by centrifugation at 7000 rpm for 20 min (Beckman Coulter Allergra X-30R, USA). The supernatant was carefully decanted when the F-MWCNTs were precipitated at the bottom of the polyethylene centrifuge tube. The procedure was repeated 8–10 times until the resistivity of the supernatant was greater than 0.5 MΩ.cm and pH $\sim$ 7.0. Finally, F-MWCNTs were rinsed with ethanol, dried overnight in a vacuum oven at 100 °C, and stored in desiccators as a dry powder for further uses.

5.3.3.1 Determination of F/O-MWCNT Water Solubility

In order to determine better MWCNT solubility, here we compared $KMnO_4$ O-MWCNTs [14] and F-MWCNTs. About 5.0 mg of either O-MWCNTs or F-MWCNTs was mixed separately into 5.0 mL of 50 mM HEPES buffer (pH: 8.0). The solutions were sonicated for 1 h to get stable homogenous colloidal suspensions. After the solutions were settled, supernatants at desired volumes were withdrawn at 0 and 30 days incubations, and the concentrations of O-MWCNTs and F-MWCNTs were determined spectrophotometrically at 500 nm [1].

5.3.4 Preparation of Nanobiohybrid

Figure 5.1 shows covalent immobilization of 3,4-POD enzyme onto F-MWCNTs using a two-step carbodiimide reaction [36]. At first, 1.0 mg of F-MWCNTs was mixed into 1.0 mL MES buffer (50 mM, pH 5.95). The resulting solution was sonicated for 30 min in order to get a homogenous well-dispersed F-MWCNT solution. A 2.0 mL solution of NHS and EDC was prepared in the same buffer and added into 1.0 mL dispersed F-MWCNT solution. The final solution (3.0 mL) was then stirred for 30 min at 400 rpm at 25 ± 1 °C. The NHS and EDC activated F-MWCNTs were then centrifuged at 15,000 rpm for 10 min in order to remove excess reactants followed by rinsing thoroughly with MES fresh buffer solution (50 mM, pH 5.95). Second, activated fresh F-MWCNTs were then transferred into a fresh and cold HEPES buffer (50 mM, pH 8.0) containing desired concentration of 3,4-POD enzyme. The cold mixture was then allowed to react under continuous magnetic stirring at 400 rpm, 4 ± 1 °C. Unbound 3,4-POD was removed by repetitive washings by fresh cold 50 mM HEPES buffer (pH 7.4) and centrifuged at 15,000 rpm for 20 min, 5 °C. Washing protocols were continued (6–7 times) until no residual enzyme activity was detected in the washing solutions, and finally once with 0.5% Tween 20 for removing nonspecific bound enzymes from F-MWCNTs.

5.3.4.1 Optimization of Nanobiohybrid Preparation

In order to get maximum 3,4-POD loading onto F-MWCNT surfaces, we optimized three important immobilization reaction conditions such as NHS: EDC, incubation times, and 3,4-POD concentrations. First, the effects of different concentrations of NHS (100, 200, 300, 600 and 1000 mM) and EDC (10, 20, 30, 60, and 100 mM) at a ratio of 3:1, and incubation times (1, 2, 4, 6 and 10 h) were checked to load 100 μg of 3,4-POD onto F-MWCNTs. Second, different concentrations of 3,4-POD (20, 40, 60, 100, 200, 500, 1000 and 2000 μg) were analyzed using optimized

NHS:EDC (mM) and time incubation (h) to get maximum 3,4-POD loading and optimum activity. The immobilization yields were measured as depicted in Eq. 5.1.

$$\text{Immobilization yield } (\%) = [(C_i - C_f)/C_i] \times 100 \qquad (5.1)$$

where C_i is the initial enzyme concentration (mg) added to the solution and C_f is the final enzyme concentration (mg) evaluated by the Bradford and BCA assays in washing solutions. All other procedures such as buffer types, pH, sonication, stirring, and so on were similar as described in the preceding Sect. 5.3.4.

5.3.4.2 Determination of 3,4-POD Concentrations

For quantifying 3,4-POD immobilized onto F-MWCNTs, all residual washing solutions after 3,4-POD immobilizations were collected and analyzed using standard BCA [6, 60] assays. Herein, BSA was used as the standard for measuring 3,4-POD enzyme concentrations.

5.3.4.3 Nanobiohybrid Activity Assays

Nanobiohybrid weight activities were determined spectrophotometrically from the disappearance of 3,4-DHBA with time at 290 nm (ε_{290} nm: 3890 M^{-1} cm^{-1}) [61, 73, 74]. The reaction cuvette contained 20; and 400 μM of 3,4-DHBA as substrates for NHS/EDC and time incubation; and enzyme loading assays, respectively. Predetermined amounts of Nanobiohybrid and buffer (50 mM HEPES, pH 7.4) were added to bring the total volume of 3.0 ml. A controlled constant stirring speed was used to ensure through mixing during the assays. Nanobiohybrid activity was expressed as the activity unit (U_x) per milligram of F-MWCNTs (W_{MWCNT}), where U_X is the activity (unit) of the immobilized 3,4-POD assayed as shown in Eq. 5.2.

$$\text{Activity (units/mg)} = U_X/W_{MWCNT} \qquad (5.2)$$

5.4 Results

5.4.1 *Characterization of F-MWCNT*

Figure 5.6 (A) shows ATR-IR spectroscopy of pristine (a) and F-MWCNTs (b). Pristine MWCNTs employed prominent peaks at 1095, 3462, 3743, and 3844 cm^{-1} (Fig. 5.6 (Aa)), suggesting the presence of different forms of –OH groups [21, 43, 68]. Some new peaks at 856, 1702, 1744, and 1794 cm^{-1} were observed in F-MWCNT (Fig. 5.6 (Ab)) as compared with pristine MWCNT (Fig. 5.6 (Aa)), attributing the presence of C-O-O-C, COOH; and C=O [12, 35, 37]. However,

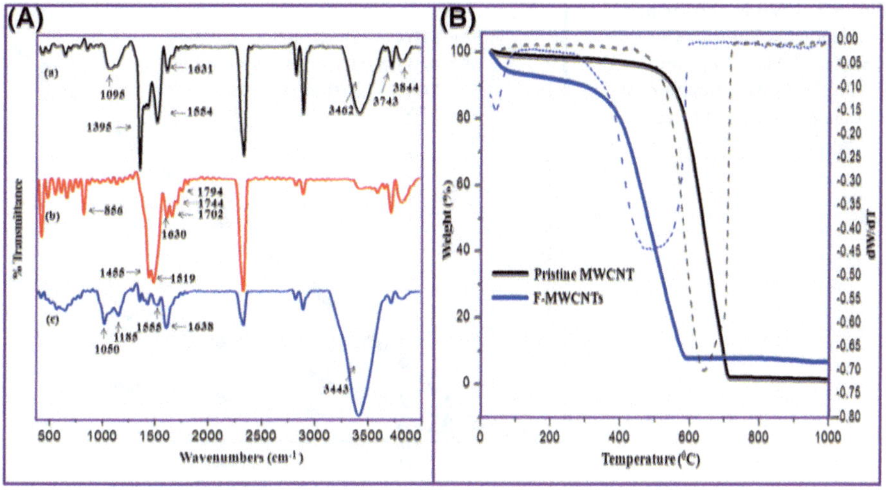

Fig. 5.6 **A** ATR-IR spectra of (*a*) pristine, (*b*) F-MWCNTs and (*c*) Nanobiohybrid; **B** TGA (*solid lines*) and DTG (*dashed lines*) plots of pristine and F-MWCNTs, respectively

some exclusive high intense and chemical shifted peaks at 1395, 1554, 1631 cm^{-1} in pristine (Fig. 5.6 (Aa)) and 1455, 1519 and 1630 cm^{-1} in F-MWCNTs (Fig. 5.6 (Ab)) were noticed because of stretching vibration of nanotube' aromatic rings [20, 71].

Figure 5.6 (B) represents TGA and DTG analyses of pristine and F-MWCNTs. Compared with pristine MWCNTs, F-MWCNTs started first stage decomposition at 100 °C because of pyrolytic evolution of water molecules [18], second weight loss occurred from 280 to 380 °C, attributing to the evaporation of –COOH groups from F-MWCNTs [64]. However, pristine MWCNTs were burned at 700 °C compared with F-MWCNTs at 580 °C, suggesting purified and functionalized nanotube' graphite sheets [23, 53].

The water solubility and colloidal stability of $KMnO_4$ O-MWCNTs and F-MWCNTs were checked by UV/vis spectroscopy as shown in Fig. 5.7. F-MWCNTs showed linear superimposable relationships between the absorbance and F-MWCNT concentrations compared with $KMnO_4$ O-MWCNTs.

5.4.2 *Characterization of Nanobiohybrid*

Figure 5.6 (Ac) shows IR characteristic bands of Nanobiohybrid catalyst. Comparing pristine- (Fig. 5.6 (Aa)) and F-MWCNTs (Fig. 5.6 (Ab)), unique IR peaks at 1050 and 1185 cm^{-1} could be resulted from C-N stretches (Coates [12]). At the same time, two major 3,4-POD IR characteristic bands at 1555 and 1638 cm^{-1} were observed in Nanobiohybrid (Fig. 5.6 (Ac)) for amide II (a blending of major N-H and minor C-N)

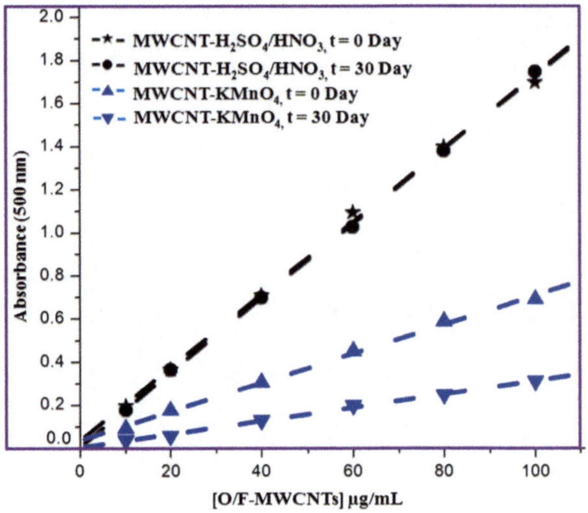

Fig. 5.7 UV/vis absorbance of aqueous solutions of $KMnO_4$ O-MWCNTs and F-MWCNTs as a function of its concentrations at 0 and 30 day intervals, respectively

and amide I (a combination of major C=O and minor C-N) stretching vibrations of its peptide backbones [34]. However, a non-negligible high intense peak at 3443 cm^{-1} (Fig. 5.6 (Ac)) might be resolved from either –NH or –OH stretching vibrations [12, 34] of 3,4-POD amino acid side chains.

Figure 5.8 shows the FE-SEM (A) and TEM (B) observations of F-MWCNTs (a) and Nanobiohybrid (b). According to Fig. 5.8 (Aa and Ba), F-MWCNTs appeared as clean, fresh, and complete amorphous carbon and metals free thread like MWCNTs. Compared with F-MWCNT (Fig. 5.8 (Aa and Ba)), images in Fig. 5.8 (Ab and Bb) clearly indicated successful immobilization of 3,4-POD onto F-MWCNT surfaces. Figure 5.8 (Ab (inset)) reveals the local binding domain of F-MWCNTs to 3,4-POD, since most of enzymes were not occupied globally (Fig. 5.8 (Ab and Bb)).

Figure 5.8 (C) shows AFM 2D (a), 3D (b) and surface roughness (c) observations of Nanobiohybrid. As shown in Fig. 5.8 (Cab), the significant difference of the Nanobiohybrid's surface structure was visualized. Controlled area where 3,4-POD was not cognizance showed smooth mean surface roughness 22 $\pm$ 8 nm (Fig. 5.8 (Cac (Profile I))) compared with three noticeable hill areas of 77 $\pm$ 28 nm in 3,4-POD heights (Fig. 5.8 (Cac (Profiles II, III and IV))).

Furthermore, Fig. 5.8 (D) represents the UV/vis spectra of F-MWCNTs, free and Nanobiohybrid solutions. It shows F-MWCNTs exhibited flat line, whereas Nanobiohybrid presented a peak at 280 nm as similar to free 3,4-POD.

Figure 5.9 shows CD spectra of control buffers (a); and free 3,4-POD and Nanobiohybrid (b). In order to quantify accurate structural changes upon 3,4-POD immobilization, we initially used F-MWCNT solution as control for Nanobiohybrid. As shown in Fig. 5.9 (a), no significant differences were observed in the entire wavelengths of F-MWCNT and CHES buffer spectra. Figure 5.9 (b) shows the differences in secondary structures (%) such as α-helices, β-sheets, turns and random

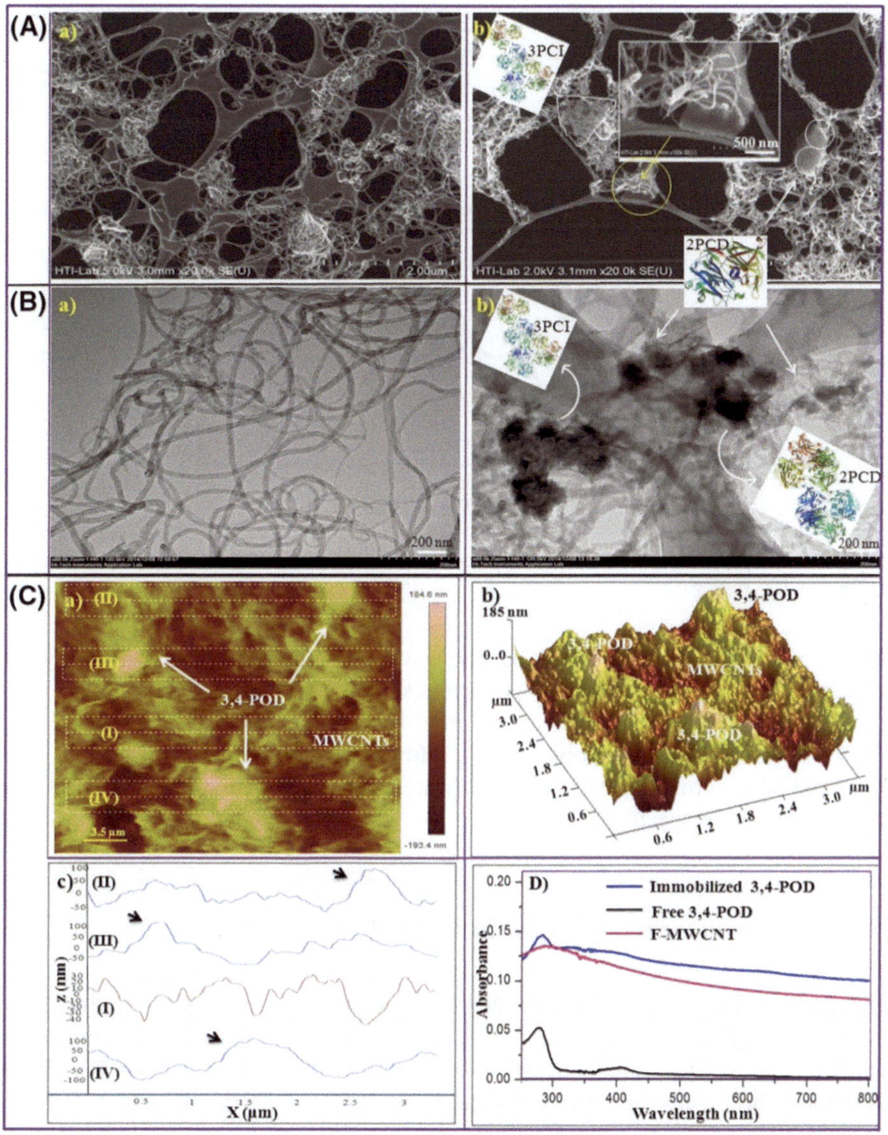

Fig. 5.8 **A** SEM and **B** TEM images of (*a*) F-MWCNTs and (*b*) Nanobiohybrid deposited on lacey copper grids; **C** AFM images of Nanobiohybrid: (*a*) bidimensional, (*b*) tridimensional analyses, herein *dotted lines* from I to IV in panel (*a*) were the sections for further analyses as shown in image profiles (*c*): *blacked arrows* indicate peaked sizes of Nanobiohybrid; and **D** UV/vis spectra of F-MWCNTs, free and Nanobiohybrid solutions

coils of free 3,4-POD and Nanobiohybrid, which were analyzed by using free online database available at http://perry.freeshell.org/raussens.html [3, 55]. The mean residue ellipticity (θ_{218}) was used to determine relative structural changes (%) of

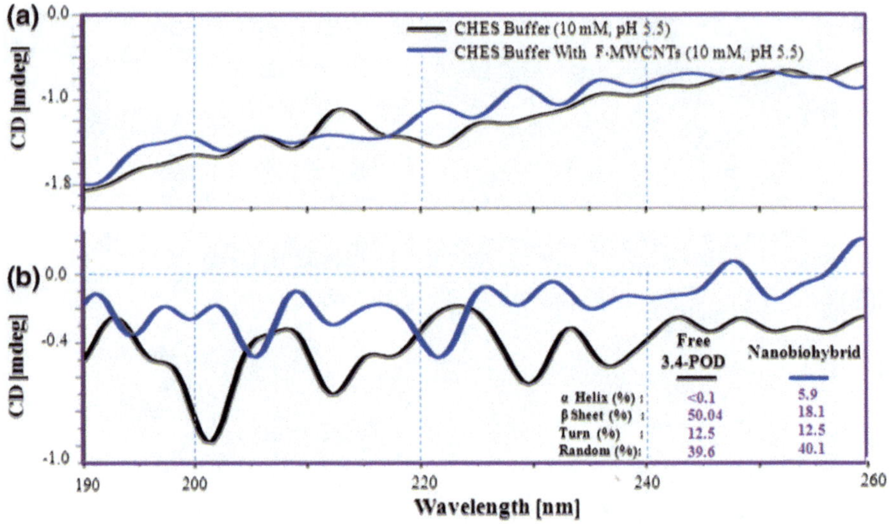

Fig. 5.9 CD spectra of (*a*) control and (*b*) sample solutions of free 3,4-POD and Nanobiohybrid in the Far-UV region

Nanobiohybrid to its free 3,4-POD and gave a reading of 44%. Conversely, we observed that α-helix content was increased from <0.1 (free 3,4-POD) to 5.9% (Nanobiohybrid), whereas a decrease in *β*-sheet was observed from 50.04 (free 3,4-POD) to 18.1% (Nanobiohybrid). However, no significant changes were observed for turn (12.5%) and random coils (40%) of free 3,4-POD and Nanobiohybrid, respectively.

5.4.3 Optimization of Nanobiohybrid

Figure 5.10 shows the effects of NHS and EDC (3:1) and time incubation on Nanobiohybrid activity (a); and the impacts of maximum 3,4-POD loading on the immobilization yield and Nanobiohybrid activity (b). According to Fig. 5.10a, Nanobiohybrid activity was increased from 0.11to 0.16 U/mg using 100 and 10 mM NHS and EDC concentrations, respectively, after which the activities were saturated in presence of its higher concentrations. Only 2h incubation of enzyme attachment could enhance Nanobiohybrid activity, after which the activity was plateau with increasing time.

In order to obtain higher enzyme loading, we investigated optimum weight ratio of 3,4-POD (0.02–2.00 mg) to F-MWCNTs (1.0 mg) as shown in Fig. 5.10b. It shows that the overall immobilization yield was increased as escalating F-MWCNTs and 3,4-POD ratios. Nanobiohybrid activity became plateau after a ratio of 1.0, suggesting a rate limiting point after which subsequent 3,4-POD attachment could saturate Nanobiohybrid activities that were flatted as shown in Fig. 5.10b.

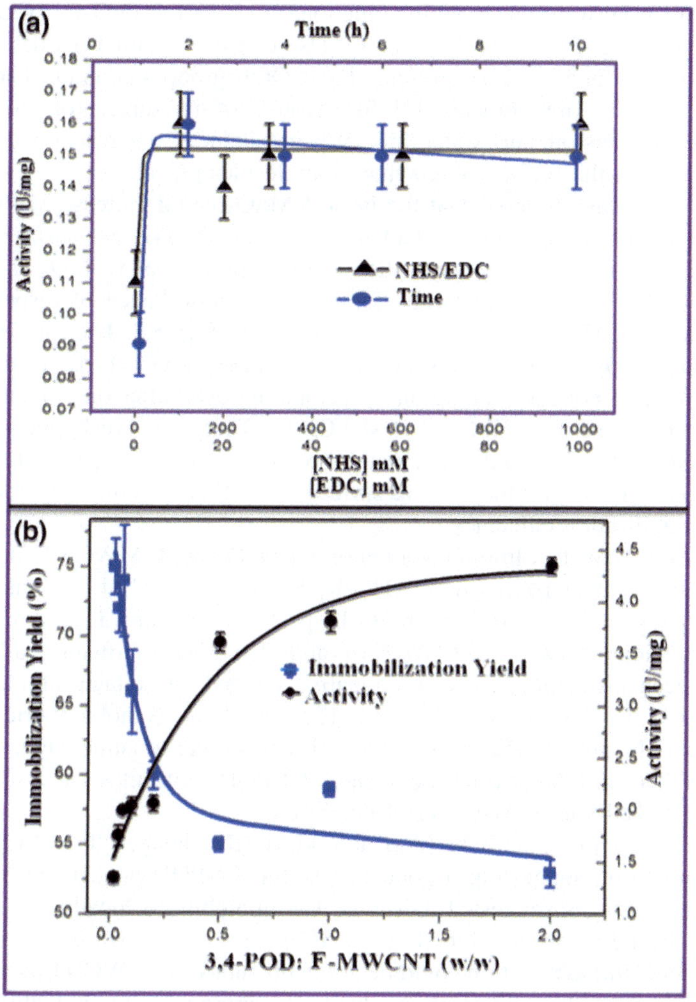

Fig. 5.10 Effects of **a** NHS and EDC concentrations, and time incubations, **b** maximum 3,4-POD enzyme loading on Nanobiohybrid activities

5.5 Discussion

Hydrophobic pristine MWCNTs were not soluble in water which results to thwart maximum enzyme loading [62]. In addition, inert CNTs required chemical modifications in order to remove non-nanotube carbon impurities and covalently attach molecules on its surface [25]. Therefore, HCl/H_2O_2 purified MWCNTs were functionalized with H_2SO_4/HNO_3 before testing enzyme immobilization. As shown in Fig. 5.6, although pristine MWCNTs (Fig. 5.6 (Aa)) indicated some OH groups, which might be generated upon atmospheric amorphous carbon oxidations [21, 43,

68], F-MWCNTs were successfully functionalized with C-O-O-C, –COOH, and –C=O groups (Fig. 5.6 (Ab)). The result is also consistent with TGA data (Fig. 5.6 (B)). Among the functionalities present, the –COOH group was an ideal attachment point for enzyme immobilization [25] because of its direct role to improve F-MWCNT dispersion into water [56, 59] and induces the reaction of the free amine groups on the surface of enzyme biomolecules [1, 32, 36].

However, it was observed that the basic $KMnO_4$ could increase MWCNT solubility by implementing a lot of –COOH groups on O-MWCNT surfaces [14]. In order to see the difference between the solubility profiles of $KMnO_4$ O-MWCNT and F-MWCNTs, we quantitatively compared the absorbance of dispersed both MWCNTs using UV/vis spectroscopy as shown in Fig. 5.7. It clearly suggested that H_2SO_4/HNO_3 treatment was better to stabilize MWCNT dispersibility into water because of not flocculating and precipitating even after one month of incubation as compared with $KMnO_4$ O-MWCNTs. Therefore, we hypothesized that H_2SO_4/HNO_3 treatment could be a method of choice to graft appropriate overhead and bring enzyme toward the reactive surface of F-MWCNTs that might lead to the better immobilization efficiency.

A successful covalent immobilization of 3,4-POD onto F-MWCNT surfaces was obtained and characterized using ATR-IR, SEM, TEM, AFM, UV/vis, and CD spectroscopy studies. Figure 5.6c shows IR peaks at 1050 and 1185 cm^{-1} for C-N bond formations between F-MWCNTs and 3,4-POD following carbodiimide reaction mechanism [36, 75]. Consequently, 3,4-POD postulated amide I and II characteristic IR bands at 1555 and 1638 cm^{-1} in Nanobiohybrid catalyst. However, C-N and amide peaks were absent in the pristine (Fig. 5.6a) and F-MWCNTs (Fig. 5.6b), which confirmed 3,4-POD molecules were successfully covalently linked with F-MWCNTs (Fig. 5.6c).

According to Fig. 5.8 SEM (Aa) and TEM (Ba) images, F-MWCNTs were smooth and fresh, suggesting a good matrix for 3,4-POD attachment. Figure 5.8 (Ab and Bb) images suggested 3,4-POD was immobilized locally, which is also supported by AFM observation (Fig. 5.8 (C)). In addition, we observed that the 3,4-POD local immobilization could be at the tip ends of F-MWCNT as depicted in insert of Fig. 5.8 (Ab). This might be due to the appropriate functionalities present at F-MWCNT tip ends [72], which might covalently interact with 3,4-POD during immobilization. The phenomenon was similar with the observation of Gao and Kyratzis [25]. The group hypothesized that the uniform distribution of enzymes onto F-MWCNTs is inconsistent with the –COOH groups that could only form at MWCNT tip ends or defect sites. Therefore, we anticipated that such local attachment could increase center to center distance of 3,4-POD which could help to avoid close packing effects and other unwanted interactions leading to high enzyme activity. It was also interesting that the 3,4-POD that was immobilized on F-MWCNTs exhibited similar shapes with the quaternary structures of it's at PDB [48, 50] as shown in Fig. 5.8 (Ab and Bb). This might be due to the 3,4-POD consists of large aggregate of different protomers [50] and each protomer could create different symmetry which depends on the enzyme sources and catalytic requirements [67].

CD is an analytical and widely used tool for determining the secondary structures of enzymes and/or proteins in solution [4]. More recent uses of CD were observed for characterizing various Nanobiohybrid catalysts and studied in great details [10, 26]. A relative structural change (44%) of Nanobiohybrid to its free 3,4-POD was anticipated and almost similar to the earlier studies performed for other Nanobiohybrid structures [10, 26]. This change along with other expected gain and loss of main secondary structures such as α-helices and β-sheets, respectively of Nanobiohybrid suggest that the hybrid catalyst was still functioning, but could decrease the overall catalytic efficiency compared with free 3,4-POD. The conformational change of Nanobiohybrid was because of bonding of free amine groups on the surface of 3,4-POD with –COOH groups of CNT [1, 10, 32, 36].

Optimization of Nanobiohybrid is a pivotal step in enzyme immobilization. The use of EDC cross-linker is well documented for durable attachment of various enzymes onto MWCNTs [25]. As a result, Gao and Kyratzis [25] critically evaluated the effects of EDC in enzyme immobilizations on CNTs [25]. The group hypothesized, since most of enzymes could immobilize on CNTs using their generic hydrophobic, electrostatic and adsorption phenomena, so far no control study was adopted to see the actual role of EDC concentrations for improving enzyme immobilization. Therefore, we evaluated the effects of EDC concentrations on 3,4-POD immobilization efficiency as shown in Fig. 5.10a. It was shown that 45% enzyme activity could be increased using only 10 mM of EDC concentration as compared with control (0.0 mM), but further increased of EDC could not affect the enzymatic activity. This might be due to the conversion of all F-MWCNT(-COOH) groups into (*O*)-acylisourea (Fig. 5.1). Since the (*O*)-acylisourea intermediate was not stable and susceptible to hydrolysis, we used NHS for yielding a semi-stable NHS ester. However, more study is necessary to re-establish the fields and be treated with caution. Figure 5.10a represents the effects of time incubation on 3,4-POD loading onto F-MWCNTs, and we observed that only a 2h incubation time was sufficient to get consistent Nanobiohybrid activity, which was much lower than that of porous glass [73, 74] and agarose matrices [76].

According to Fig. 5.10b, maximum (75) and minimum (53%) immobilization yields were observed when lowest and highest 3,4-POD concentrations were used, respectively. The loading of 1060 μg of 3,4-POD/mg of MWCNT (53%) was still considered to be the highest compared with other Nanobiohybrids [1, 38, 51, 63, 78]. The yield might be due to the use of highly purified and water soluble F-MWCNT matrix as support, optimized immobilization reaction conditions, and the nature of enzyme molecules. However, Nanobiohybrid activity became plateau after a ratio of 1.0 which helped to get maximum catalytic activity using a minimal enzyme waste for subsequent experimental procedures.

5.6 Conclusions

This chapter demonstrated to prepare a suitable F-MWCNT matrix for 3,4-POD immobilization. Although two wet oxidizing agents such as $KMnO_4$ and H_2SO_4/HNO_3 could anchor –COOH groups on MWCNTs, H_2SO_4/HNO_3 attracted more attention and played better role in achieving stable MWCNT colloidal solution even for a month. Convincing evidences from the SEM, TEM, and AFM along with UV/vis and IR spectroscopic data clearly suggested that the 3,4-POD was successfully immobilized onto F-MWCNTs. CD spectra showed 44% conformational changes of Nanobiohybrid to its free 3,4-POD structure, which might indicate to have sufficient catalytic efficiency for 3,4-DHBA binding. By optimizing important enzyme immobilization reaction conditions such as the use cross-linker, time incubation and different concentrations of 3,4-POD loading helped us to get the maximum enzyme loading (1060 μg of 3,4-POD/mg of MWCNT). This paves the way for the development of effective Nanobiohybrid that might have the imminent potentiality to purify 3,4-DHBA contaminated wastewater.

References and Future Readings

1. Asuri, P., Karajanagi, S.S., Sellitto, E., Kim, D.Y., Kane, R.S., Dordick, J.S.: Water-soluble carbon nanotube-enzyme conjugates as functional biocatalytic formulations. Biotechnol. Bioeng. **95**(5), 804–811 (2006)
2. Babich, H., Sedletcaia, A., Kenigsberg, B.: In vitro cytotoxicity of protocatechuic acid to cultured human cells from oral tissue: Involvement in oxidative stress. Pharmacol. Toxicol. **91** (5), 245–253 (2002)
3. Bayat, S., Tejo, B.A., Abdulmalek, E., Salleh, A.B., Normi, Y.M., Rahman, M.B.A.: Rational design of mimetic peptides based on aldo-ketoreductase enzyme as asymmetric organocatalysts in aldol reactions. RSC Advances **4**(73), 38859–38868 (2014)
4. Berova, N., Nakanishi, K., Woody, R.: Circular Dichroism: Principles and Applications. Wiley, US (2000)
5. Blanchard, N., Hatton, R., Silva, S.: Tuning the work function of surface oxidised multi-wall carbon nanotubes via cation exchange. Chem. Phys. Lett. **434**(1), 92–95 (2007)
6. Bradford, M.M.: A rapid and sensitive method for the quantitation of microgram quantities of protein utilizing the principle of protein-dye binding. Anal. Biochem. **72**(1), 248–254 (1976)
7. Brena, B., González-Pombo, P., Batista-Viera, F.: Immobilization of Enzymes: A Literature Survey. In: Immobilization of Enzymes and Cells, pp 15–31. Springer, Berlin (2013)
8. Bubinas, A., Giedraitytė, G., Kalėdienė, L.: Protocatechuate 3, 4-dioxygenase from thermophilic Geobacillus sp. strain. Biologija **53**(1), (2007)
9. Bull, C., Ballou, D.: Purification and properties of protocatechuate 3, 4-dioxygenase from Pseudomonas putida. A new iron to subunit stoichiometry. J. Biol. Chem. **256**(24), 12673–12680 (1981)
10. Cang-Rong, J.T., Pastorin, G.: The influence of carbon nanotubes on enzyme activity and structure: Investigation of different immobilization procedures through enzyme kinetics and circular dichroism studies. Nanotechnology **20**(25), 255102 (2009)
11. Chen, Y., Dilworth, M., Glenn, A.: Aromatic metabolism in Rhizobium trifolii—protocatechuate 3, 4-dioxygenase. Arch. Microbiol. **138**(3), 187–190 (1984)

12. Coates, J.: Interpretation of Infrared Spectra, A Practical Approach. Encyclopedia of analytical chemistry (2000)
13. Corry, B.: Water and ion transport through functionalised carbon nanotubes: Implications for desalination technology. Energy Environ. Sci. **4**(3), 751–759 (2011)
14. Das, R., Abd Hamid, S.B., Ali, M.E., Annuar, M.S.M., Samsudin, E.M.B., Bagheri, S.: Covalent Functionalization Schemes for Tailoring Solubility of Multi-Walled Carbon Nanotubes in Water and Acetone Solvents. Science of Advanced Materials In Press (2015)
15. Das, R., Abd Hamid, S.B., Ali, M.E., Ismail, A.F., Annuar, M.S.M., Ramakrishna, S.: Multifunctional carbon nanotubes in water treatment: The present, past and future. Desalination **354**(0), 160–179 (2014). doi:http://dx.doi.org/10.1016/j.desal.2014.09.032
16. Das, R., Ali, M.E., Bee Abd Hamid, S., Annuar, M., Ramakrishna, S.: Common wet chemical agents for purifying multiwalled carbon nanotubes. J. Nanomaterials **2014**, 237 (2014)
17. Das, R., Ali, M.E., Bee Abd Hamid, S., Annuar, M., Ramakrishna, S.: Multifunctional carbon nanotubes in water treatment: The present, past and future. Desalination **354**,160–179 (2014)
18. Datsyuk, V., Kalyva, M., Papagelis, K., Parthenios, J., Tasis, D., Siokou, A., Kallitsis, I., Galiotis, C.: Chemical oxidation of multiwalled carbon nanotubes. Carbon **46**(6), 833–840 (2008)
19. Di Nardo, G., Roggero, C., Campolongo, S., Valetti, F., Trotta, F., Gilardi, G.: Catalytic properties of catechol 1, 2-dioxygenase from Acinetobacter radioresistens S13 immobilized on nanosponges. Dalton Trans. **33**, 6507–6512 (2009)
20. Dos Reis, G., Dias, I., De Santana, H., Duarte, J., Laureto, E., Di Mauro, E., Da Silva, M.: Analysis of optical properties of poly (3-methylthiophene)(P3MT) electrochemically synthesized. Synth. Met. **161**(3), 340–347 (2011)
21. Esteve, W., Budzinski, H., Villenave, E.: Relative rate constants for the heterogeneous reactions of OH, NO_2 and NO radicals with polycyclic aromatic hydrocarbons adsorbed on carbonaceous particles. Part 1: PAHs adsorbed on 1–2 μm calibrated graphite particles. Atmos. Environ. **38**(35), 6063–6072 (2004)
22. Feng, W., Ji, P.: Enzymes immobilized on carbon nanotubes. Biotechnol. Adv. **29**(6), 889–895 (2011)
23. Feng, Y., Zhang, H., Hou, Y., McNicholas, T.P., Yuan, D., Yang, S., Ding, L., Feng, W., Liu, J.: Room temperature purification of few-walled carbon nanotubes with high yield. ACS Nano. **2**(8), 1634–1638 (2008)
24. Fujisawa, H., Hayaishi, O.: Protocatechuate 3, 4-dioxygenase I. Crystallization and characterization. J. Biol. Chem. **243**(10), 2673–2681 (1968)
25. Gao, Y., Kyratzis, I.: Covalent immobilization of proteins on carbon nanotubes using the cross-linker 1-ethyl-3-(3-dimethylaminopropyl) carbodiimide—A critical assessment. Bioconjugate Chem. **19**(10), 1945–1950 (2008)
26. Goh, W.J., Makam, V.S., Hu, J., Kang, L., Zheng, M., Yoong, S.L., Udalagama, C.N., Pastorin, G.: Iron oxide filled magnetic carbon nanotube-enzyme conjugates for recycling of amyloglucosidase: Toward useful applications in biofuel production process. Langmuir **28** (49), 16864–16873 (2012)
27. Grady, Jr. C.L., Daigger, G.T., Love, N.G., Filipe, C.D., Leslie Grady, C.: Biological wastewater treatment, vol. 3. IWA Publishing (2011)
28. Guzik, U., Hupert-Kocurek, K., Krysiak, M., Wojcieszyńska, D.: Degradation Potential of Protocatechuate 3, 4-Dioxygenase from Crude Extract of Stenotrophomonas maltophilia Strain KB2 Immobilized in Calcium Alginate Hydrogels and on Glyoxyl Agarose. BioMed research international **2014** (2014)
29. Guzik, U., Hupert-Kocurek, K., Wojcieszyńska, D.: Intradiol dioxygenases—The key enzymes in xenobiotics degradation (2013)
30. Hammer, A., Stolz, A., Knackmuss, H.-J.: Purification and characterization of a novel type of protocatechuate 3, 4-dioxygenase with the ability to oxidize 4-sulfocatechol. Arch. Microbiol. **166**(2), 92–100 (1996)

31. Hemmrich, K., Salber, J., Meersch, M., Wiesemann, U., Gries, T., Pallua, N., Klee, D.: Three-dimensional nonwoven scaffolds from a novel biodegradable poly (ester amide) for tissue engineering applications. J. Mater. Sci.—Mater. Med. **19**(1), 257–267 (2008)
32. Huang, W., Taylor, S., Fu, K., Lin, Y., Zhang, D., Hanks, T.W., Rao, A.M., Sun, Y.-P.: Attaching proteins to carbon nanotubes via diimide-activated amidation. Nano Lett. **2**(4), 311–314 (2002)
33. Hwang, E.T., Gu, M.B.: Enzyme stabilization by nano/microsized hybrid materials. Eng. Life Sci. **13**(1), 49–61 (2013)
34. Jabs, A.: Determination of secondary structure in proteins by fourier transform infrared spectroscopy (FTIR). From the IMB Jena Image Library web page http://imb-jenade/ImgLibDoc/ftir/IMAGE_FTIR.html (2005). Accessed 1 Aug 2005
35. Jarmelo, S., Reva, I., Lapinski, L., Nowak, M., Fausto, R.: Matrix-isolated diglycolic anhydride: Vibrational spectra and photochemical reactivity. J. Phys. Chem. A **112**(44), 11178–11189 (2008)
36. Jiang, K., Schadler, L.S., Siegel, R.W., Zhang, X., Zhang, H., Terrones, M.: Protein immobilization on carbon nanotubes via a two-step process of diimide-activated amidation. J. Mater. Chem. **14**(1), 37–39 (2004)
37. Jonoobi, M., Mathew, A.P., Abdi, M.M., Makinejad, M.D., Oksman, K.: A comparison of modified and unmodified cellulose nanofiber reinforced polylactic acid (PLA) prepared by twin screw extrusion. J. Polym. Environ. **20**(4), 991–997 (2012)
38. Karajanagi, S.S., Vertegel, A.A., Kane, R.S., Dordick, J.S.: Structure and function of enzymes adsorbed onto single-walled carbon nanotubes. Langmuir **20**(26), 11594–11599 (2004)
39. Khin, M.M., Nair, A.S., Babu, V.J., Murugan, R., Ramakrishna, S.: A review on nanomaterials for environmental remediation. Energy Environ. Sci. **5**(8), 8075–8109 (2012)
40. Koppisch, A.T., Fox, D.T., Hotta, K., Welsh, J.D.: Production of industrially relevant compounds in prokaryotic organisms. Google Patents (2011)
41. Li, J-h, R-y, Hong, G-h, Luo, Zheng, Y., H-z, Li, D-g, Wei: An easy approach to encapsulating Fe_3O_4 nanoparticles in multiwalled carbon nanotubes. New Carbon Mater. **25** (3), 192–198 (2010)
42. Liu, Y., Qu, X., Guo, H., Chen, H., Liu, B., Dong, S.: Facile preparation of amperometric laccase biosensor with multifunction based on the matrix of carbon nanotubes–chitosan composite. Biosens. Bioelectron. **21**(12), 2195–2201 (2006)
43. Ma, P.-C., Kim. J.-K.: Carbon nanotubes for polymer reinforcement. CRC Press (2011)
44. Matsuura, K., Saito, T., Okazaki, T., Ohshima, S., Yumura, M., Iijima, S.: Selectivity of water-soluble proteins in single-walled carbon nanotube dispersions. Chem. Phys. Lett. **429** (4), 497–502 (2006)
45. Munge, B., Guodong, L., Greg, C., Joseph, W.: Multiple enzyme layers on carbon nanotubes for electrochemical detection down to 80 DNA copies. Anal. Chem. **77**(14), 4662–4666 (2005)
46. Mugdha, A., Usha, M.: Enzymatic Treatment of wastewater containing dyestuffs using different delivery systems. Sci. Rev. Chem. Communi. **2**(1), 31–40, (2012). ISSN 2277 2669
47. Neun, D.J., Penn, A., Snyder, C.A.: Evidence for strain-specific differences in benzene toxicity as a function of host target cell susceptibility. Arch. Toxicol. **66**(1), 11–17 (1992)
48. Ohlendorf, D.H., Orville, A.M., Lipscomb, J.D.: Structure of Protocatechuate 3, 4-Dioxygenase from Pseudomonas aeruginosa at 2.15 Å Resolution. J. Mol. Biol. **244**(5), 586–608 (1994)
49. Ornston, L.: The conversion of Catechol and Protocatechuate to β-Ketoadipate by Pseudomonas putida II. Enzymes of the protocatechuate pathway. J. Biol. Chem. **241**(16), 3787–3794 (1966)
50. Orville, A.M., Elango, N., Lipscomb, J.D., Ohlendorf, D.H.: Structures of competitive inhibitor complexes of protocatechuate 3, 4-dioxygenase: Multiple exogenous ligand binding orientations within the active site. Biochemistry **36**(33), 10039–10051 (1997)

51. Pavlidis, I.V., Vorhaben, T., Tsoufis, T., Rudolf, P., Bornscheuer, U.T., Gournis, D., Stamatis, H.: Development of effective nanobiocatalytic systems through the immobilization of hydrolases on functionalized carbon-based nanomaterials. Bioresour. Technol. **115**, 164–171 (2012)
52. Pedrosa, V.A., Paliwal, S., Balasubramanian, S., Nepal, D., Davis, V., Wild, J., Ramanculov, E., Simonian, A.: Enhanced stability of enzyme organophosphate hydrolase interfaced on the carbon nanotubes. Colloids Surf. B: Biointerfaces **77**(1), 69–74 (2010)
53. Peng, Y., Liu, H.: Effects of oxidation by hydrogen peroxide on the structures of multiwalled carbon nanotubes. Ind. Eng. Chem. Res. **45**(19), 6483–6488 (2006)
54. Qiu, Y., Park, K.: Environment-sensitive hydrogels for drug delivery. Adv. Drug Deliv. Rev. **53**(3), 321–339 (2001)
55. Raussens, V., Ruysschaert, J.-M., Goormaghtigh, E.: Protein concentration is not an absolute prerequisite for the determination of secondary structure from circular dichroism spectra: A new scaling method. Anal. Biochem. **319**(1), 114–121 (2003)
56. Shieh, Y.-T., Liu, G.-L., Wu, H.-H., Lee, C.-C.: Effects of polarity and pH on the solubility of acid-treated carbon nanotubes in different media. Carbon **45**(9), 1880–1890 (2007)
57. Shimizu, A., Tanaka, K., Fujimori, M.: Abatement technologies for N_2O emissions in the adipic acid industry. Chemosphere-global Change Sci **2**(3), 425–434 (2000)
58. Silva, ASd, Jacques, R.J.S., Andreazza, R., Bento, F.M., Camargo, FAdO: The effects of trace elements, cations, and environmental conditions on protocatechuate 3, 4-dioxygenase activity. Scientia Agricola **70**(2), 68–73 (2013)
59. Smith, B., Wepasnick, K., Schrote, K.E., Cho, H.-H., Ball, W.P., Fairbrother, D.H.: Influence of surface oxides on the colloidal stability of multi-walled carbon nanotubes: A structure—property relationship. Langmuir **25**(17), 9767–9776 (2009)
60. Smith, P., Krohn, R.I., Hermanson, G., Mallia, A., Gartner, F., Provenzano, M., Fujimoto, E., Goeke, N., Olson, B., Klenk, D.: Measurement of protein using bicinchoninic acid. Anal. Biochem. **150**(1), 76–85 (1985)
61. Stanier, R., Ingraham, J.: Protocatechuic acid oxidase. J. Biol. Chem. **210**(2), 799–808 (1954)
62. Subrizi, F., Crucianelli, M., Grossi, V., Passacantando, M., Pesci, L., Saladino, R.: Carbon nanotubes as activating tyrosinase supports for the selective synthesis of catechols. ACS Catalysis **4**(3), 810–822 (2014)
63. Suma, Y., Kim, D., Lee, J.W., Park, K.Y., Kim, H.S.: Degradation of catechol by immobilized hydroxyquinol 1, 2-dioxygenase (1, 2-HQD) onto single-walled carbon nanotubes. In: Proceedings of the International Conference on Chemical, Environmental Science and Engineering (ICEEBS'12) (2012)
64. Tang, M., Dou, H., Sun, K.: One-step synthesis of dextran-based stable nanoparticles assisted by self-assembly. Polymer **47**(2), 728–734 (2006). doi:10.1016/j.polymer.2005.11.091
65. Tran, D.N., Balkus Jr., K.J.: Perspective of recent progress in immobilization of enzymes. Acs Catalysis **1**(8), 956–968 (2011)
66. Verma, M.L., Barrow, C.J., Puri, M.: Nanobiotechnology as a novel paradigm for enzyme immobilisation and stabilisation with potential applications in biodiesel production. Appl. Microbiol. Biotechnol. **97**(1), 23–39 (2013)
67. Vetting, M.W., D'Argenio, D.A., Ornston, L.N., Ohlendorf, D.H.: Structure of Acinetobacter strain ADP1 protocatechuate 3, 4-dioxygenase at 2.2 Å resolution: Implications for the mechanism of an intradiol dioxygenase. Biochemistry **39**(27), 7943–7955 (2000)
68. Vione, D., Maurino, V., Minero, C., Pelizzetti, E., Harrison, M.A., Olariu, R.-I., Arsene, C.: Photochemical reactions in the tropospheric aqueous phase and on particulate matter. Chem. Soc. Rev. **35**(5), 441–453 (2006)
69. Wagner, M., Loy, A., Nogueira, R., Purkhold, U., Lee, N., Daims, H.: Microbial community composition and function in wastewater treatment plants. Antonie Van Leeuwenhoek **81**(1–4), 665–680 (2002)
70. Wandrey, C., Liese, A., Kihumbu, D.: Industrial biocatalysis: Past, present, and future. Org. Process Res. Dev. **4**(4), 286–290 (2000)

71. Wang, W.-S., Wang, D.-H., Qu, W.-G., Lu, L.-Q., Xu, A.-W.: Large ultrathin anatase TiO_2 nanosheets with exposed 001 facets on graphene for enhanced visible light photocatalytic activity. J. Phys. Chem. C **116**(37), 19893–19901 (2012)
72. Wepasnick, K.A., Smith, B.A., Schrote, K.E., Wilson, H.K., Diegelmann, S.R., Fairbrother, D.H.: Surface and structural characterization of multi-walled carbon nanotubes following different oxidative treatments. Carbon **49**(1), 24–36 (2011)
73. Wojtaś-Wasilewska, M., Luterek, J., Rogalski, J.: Immobilization of protocatechuate 3, 4-dioxygenase from Pleurotus ostreatus on activated porous glass beads. Phytochemistry **27** (9), 2731–2733 (1988)
74. Wojtaś-Wasilewska, M., Luterek, J., Leonowicz, A., Dawidowicz, A.: Dearomatization of lignin derivatives by fungal protocatechuate 3, 4-dioxygenase immobilized on porosity glass. Biotechnol. Bioeng. **32**(4), 507–511 (1988)
75. Xu, R., Chi, C., Li, F., Zhang, B.: Laccase-Polyacrylonitrile nanofibrous membrane: Highly immobilized, stable, reusable, and efficacious for 2, 4, 6-Trichlorophenol removal. ACS Appl. Mater. Interfaces. **5**(23), 12554–12560 (2013)
76. Zaborsky, O.R., Ogletree, J.: Immobilization of protocatechuate 3, 4-dioxygenase with activated agarose. Biochimica et Biophysica Acta (BBA)-Enzymology **289**(1), 68–76 (1972)
77. Zaks, A.: Industrial biocatalysis. Curr. Opin. Chem. Biol. **5**(2), 130–136 (2001)
78. Zhai, R., Zhang, B., Wan, Y., Li, C., Wang, J., Liu, J.: Chitosan–halloysite hybrid-nanotubes: Horseradish peroxidase immobilization and applications in phenol removal. Chem. Eng. J. **214**, 304–309 (2013)
79. Zylstra, G., Olsen, R., Ballou, D.: Genetic organization and sequence of the Pseudomonas cepacia genes for the alpha and beta subunits of protocatechuate 3, 4-dioxygenase. J. Bacteriol. **171**(11), 5915–5921 (1989)

Chapter 6
Nanobiohybrid for Water Treatment

Abstract Nanobiohybrid has recently been grown out of other water purification technologies. The present study reported for the first time to demonstrate the uses of Nanobiohybrid for the effective degradation of 3,4-DHBA in water. Compared with free 3,4-POD, Nanobiohybrid showed greater stabilities in higher alkaline pH and temperature zones. The free 3,4-POD lost its residual activity of 82%, while Nanobiohybrid was 66% after 180 min incubations at 90 °C. Moreover, Nanobiohybrid could retain 93% and about 50% of its relative activity and overall catalytic efficiency to the free 3,4-POD, respectively. Higher storage stability of the Nanobiohybrid was observed, since it maintained >55% of residual activity compared with free 3,4-POD which was almost 40% after 30 days of storage at both 4 and 25 °C. Recrudescent Nanobiohybrid could keep >60% of residual activity after ten operational cycles used, endowing to decrease the production costs of 3,4-POD for long term uses. More than 70% of 3,4-DHBA removed by the Nanobiohybrid in less than 4 h treatment, suggesting a reduced time protocol. Therefore, with these overall results analyses we can conclude that the developed Nanobiohybrid here could act as an efficient novel decontamination platform for mineralizing 3,4-DHBA in water.

6.1 Background

Industrial foods processing wastewater effluents have degraded finite fresh water bodies by escalating the concentrations of 3,4-DHBA into natural aquatic systems [21]. This has shown detrimental effects on terrestrial, aquatic, and aerial, flora and fauna. Only 1.0 μmol of 3,4-DHBA could induce oxidative stress, tumor production, and inflammation in mice cells [1, 17, 18]. A toxic dose of 3,4-DHBA (0.1% in drinking water for 60 days administration) has shown to decrease glutathione (major cellular antioxidant) and increase alanine aminotransferase levels which indicate hepatotoxicity and neurotoxicity of 3,4-DHBA on mice cells [17]. Similar effects of 3,4-DHBA have also been observed where an initial toxicity has occurred at 5.0 mM of DHBA to normal human oral tissues. Therefore, the enzyme 3,4-POD

R. Das, *Nanohybrid Catalyst based on Carbon Nanotube*,
Carbon Nanostructures, DOI 10.1007/978-3-319-58151-4_6

which could degrade this toxic 3,4-DHBA [1] to nontoxic muconic acid [19] has an appeal for developing an effective and inexpensive regulatory tool to control the quantity of 3,4-DHBA in public sewage water matrices.

Although some studies such as Fenton methods [9, 21], adsorption [22], O_3/UV or H_2O_2/UV methods [2] and microbial degradation [3] have been adopted to eradicate 3,4-DHBA from water, no study has yet been realized to degrade 3,4-DHBA using Nanobiohybrid catalytic systems. The Nanobiohybrid could offer additional advantages over other degradation processes, since it requires mild reaction conditions, has high selectivity, sensitivity, and reusability, functioning in harsh environments, no byproducts, and eco-friendly due to less chemical masses [14]. Consequently, biocatalytic water purification systems have recently been popular for removing a wide range of water pollutants such as 2,4-Dichlorophenol (2,4-DCP) [31], 2,4,6-Trichlorophenol (2,4,6-TCP) [30], organophosphates (OP) [16], and so on [8].

In this study, we used our developed Nanobiohybrid catalyst for the direct degradation of 3,4-DHBA in artificial wastewater. Results showed that the Nanobiohybrid had broader working alkaline pH, higher temperature ranges, a longer shelf life, and recrudescence and reusability in degrading 3,4-DHBA, and thereby proving a novel technology for purifying 3,4-DHBA contaminated water with reduced costs and time.

6.2 Literature Review

While conventional water treatment technologies such as distillation, filtration, sedimentation, advanced membrane technologies, sludge storage, coagulation, and flocculation, ozone, chlorine, ultraviolet, H_2O_2 oxidation, solar water disinfection or photocatalytic degradation, supercritical water oxidation, sonochemical degradation, and microbial water sludge treatments have been facing problems with instability, nonspecificity and selectivity, energy intensive, and so on [14]; Nanobiohybrid has recently been budged to provide alternative efficient way to purify wastewater effluents. For example, while photocatalysis and CWAO have been facing serious technological drawbacks such as energy intensive, toxic intermediates production, less selectivity, and sensitivity for dilute solutions and leaching effects, which have motivated us to develop a novel route of catalysis for water purification called "NanoBiohybrid Catalyst" technology. Our vision was to utilize all the superiors' physicochemical properties of functionalized CNT, which could be used as a nanocarrier for enzyme immobilization to birth "NanoBiohybrid Catalyst." Using enzyme in water purification is not new, but developing a stable, efficient, robust, and cheap enzyme based system is challenging. We claim that our developed "NanoBiohybrid Catalyst" could be used to attenuate phenolic wastewater pollutants. The catalyst has found highly stable and efficient with a high number of reusability. Interestingly, the product that has been generated from the pollutant's mineralization is a nontoxic and acts as a precursor for value-added adipic acid production: a dream “from wastewater to wealth.” As a corollary, the catalyst would also have industrial appeal. Besides, a great promise and operational feasibility may come from the development of a Nanobiohybrid Catalyst based “One Pot” water purification system. Herein different enzyme catalysts can be attached on functionalized nanocarrier for tracing a wide range of water pollutants. We believe that building of such system would have many advantages over conventional water purification technologies. First, CNT hollow structure could allow a frictionless water transport facility which might decrease the overall pressure-driven forces and ultimately energy requirement. Second, it can be used to tackle a diverse range of micro-contaminants in water. Third, the cytotoxic properties of CNTs might inhibit the system to be fouling which is the major burden for wide spreading of current membrane technologies. Fourth, the Pot can simultaneously sense and quantify both the permissible and nonpermissible limit of different micro-pollutants in drinking water supply. In the fifth, it can act as key industrial catalyst to produce value-added compounds from the degradation of wastewater organic pollutants. Finally, it can be used as point-of-use (POU) device at point-of-generation (POG) of pollutant effluents, and ultimately decreased labor and materials costs. An initial foray into this emerging research area has ensured us the satisfactory results of Nanobiohybrid published by different research groups. First, Xu et al. [30, 31]) developed a Nanobiohybrid by immobilizing laccase onto chitosan/PVA fabricated nanofiber for degrading 2,4-DCP in water [31] (Fig. 6.1). The group was successfully immobilized laccase 853 mg/g support which showed less sensitivity toward changing pH, temperatures, and oscillation. It could retain 54% of residual activity after seven-cycle of operational uses. A higher removal efficiency (RE) of Nanobiohybrid (87.6) compared with free laccase (82.7%) was observed in degrading 2,4-DCP after 6 h batch experiment.

Second, in the same year, Xu et al. [30] continued to develop another Nanobiohybrid for 2,4,6-TCP removal [30]. Herein laccase was immobilized onto

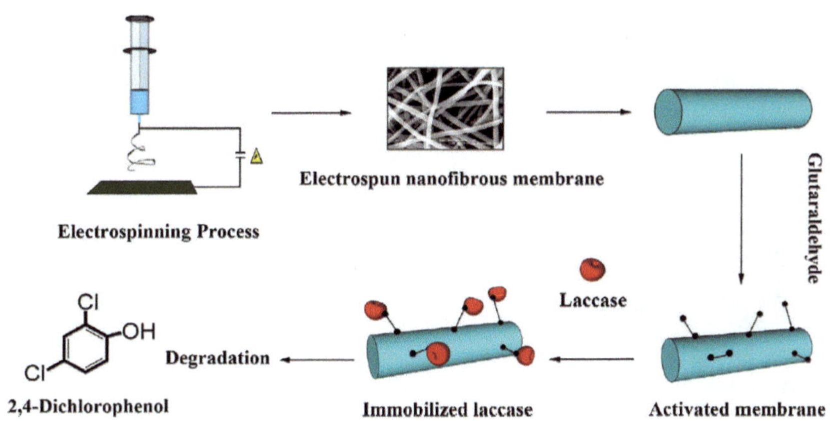

Fig. 6.1 A schematic representation for electrospinning process, laccase immobilization and 2,4-DCP degradation. Figure is adapted with permission from Elsevier [31]

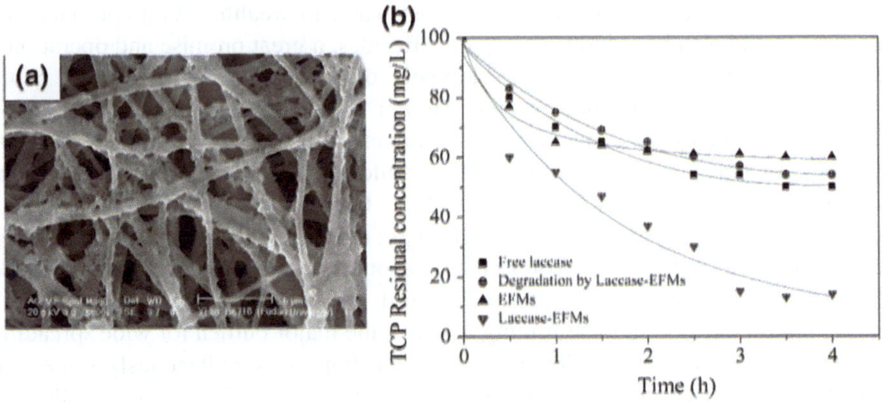

Fig. 6.2 SEM micrographs of Nanobiohybrid **a** and degradation kinetics of TCP by free and immobilized laccase **b** Figure is adapted with permission from ACS [30]

polyacrylonitrile nanofibrous membrane (Fig. 6.2a) and showed excellent satisfactory results in terms of stability, reusability, and high RE toward the 2,4,6-TCP removal (Fig. 6.2b). The maximum loading of laccase was 220 mg/g fibers that could retain 72% of relative activity to its free enzyme. A higher reusability of Nanobiohybrid was observed with residual activity of 60% after ten operational cycle uses and also kept 92% of residual activity on 18 days storage, which was only 20% for the free laccase. A higher RE (87%) of the Nanobiohybrid was obtained as compared with free laccase (50%) for degrading 2,4,6-TCP in 4 h batch experiment (Fig. 6.2b).

At last, Mechrez et al. [16] mineralized OP such as methyl paraoxon (MOX) by immobilizing organophosphate hydrolase (OPH) onto CNTs (Fig. 6.3).

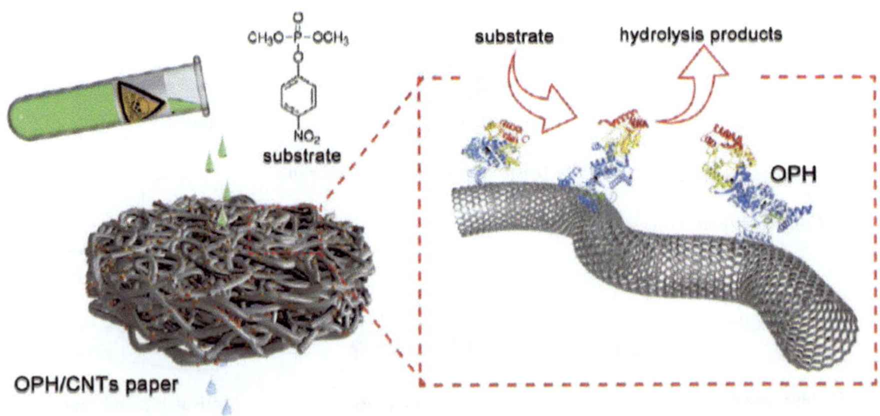

Fig. 6.3 Action of Nanobiohybrid for MOX mineralization. Figure is adapted with permission from RSC Mechrez et al. [16]

Predetermined MOX concentration (1.5 mM) was used in order to measure the potentiality of Nanobiohybrid catalyst. Within first filtration cycle it could decrease 10% of MOX, and the residual 1.16 mM was remained after four filtration cycles. Therefore, the Nanobiohybrid-mediated degradation of wastewater pollutants is of high interest in scientific community. This is because of its ease in catalytic reusability, stability, and continuous operations [5]. Using nanoscale support material for enzyme immobilization could give additional advantages for water purification such as high effluent load and less energy costs [5].

6.3 Materials and Methods

6.3.1 *Materials and Reagents*

The materials and reagents for Nanobiohybrid development were similar as described in Chap. 5, Sect. 5.3.1. Lyophilized powder of pure 3,4-POD ($\geq$ 3.0 U/mg solid) from *Pseudomonas* sp. was purchased from Sigma-Aldrich Sdn Bhd. (Malaysia) and used as free enzyme without further purifications. HCl (37%), sodium hydroxide, MES, HEPES, CHES, 4-(Cyclohexylamino)-1-butanesulfonic acid (CABS) and 3,4-DHBA were purchased from Sigma-Aldrich Sdn. Bhd. (Malaysia). Deionized water (pH 6.8) was used in all of the experiments.

6.3.2 *Free 3,4-POD and Nanobiohybrid Activity Assays*

The free 3,4-POD and Nanobiohybrid activities were determined spectrophotometrically from the disappearance of 3,4-DHBA with time at 290 nm ($\varepsilon_{290\ nm}$: 3890 M^{-1} cm^{-1}) [24]; Wojtaś-Wasilewska et al. [28, 29]. The reaction cuvette contained 100 µM of 3,4-DHBA, predetermined amounts of free 3,4-POD, and Nanobiohybrid and buffer to bring the total volume of 3.0 mL. Optimum pH of free 3,4-POD and Nanobiohybrid was determined using 50 mM of [MES, pH (4.0–7.0), HEPES (7.5–8.5), CHES (8.5–10.0), and CABS (10.5–11.0)] at 30 °C. Temperature ranges from 5 to 80 °C were studied at 50 mM CHES and optimum pH 9.0 and 10.0 for higher free 3,4-POD and Nanobiohybrid activities, respectively. A controlled constant stirring speed was used to ensure through mixing during the assay. One unit of activity was defined as the amount of enzyme required to oxidize 1.0 µmol of 3,4-DHBA per minute. Nanobiohybrid activity was expressed as activity unit (U_X) per milligram of F-MWCNTs (W_{MWCNT}), where U_X, is the activity (unit) of the immobilized 3,4-POD assayed by similar free enzyme method as Eq. (6.1).

$$\text{Activity (units/mg)} = \text{U}_\text{X}/\text{W}_\text{MWCNT} \quad (6.1)$$

The kinetic parameters (V_{max}, K_m, K_{cat}, and K_{cat}/K_m) of the free 3,4-POD and Nanobiohybrid were calculated using Michaelis–Menten model [13] as given in Eq. (6.2) with 3,4-DHBA concentrations (1–200 μM) at 50 mM CHES, optimum pH 9.0 and 10.0 with optimum temperatures 55 and 60 °C, respectively,

$$\frac{1}{v} = \frac{K_m}{V_{max}} X \frac{1}{[S]} + \frac{1}{V_{max}} \quad (6.2)$$

where V is an initial reaction rate (nmole min^{-1}); K_m is the Michaelis–Menten constant (μM); V_{max} is the maximum reaction rate (nmole min^{-1}), and [S] is an initial 3,4-DHBA concentration (μM).

6.3.3 Free 3,4-POD and Nanobiohybrid Stability Assays

Temperature stabilities of the free 3,4-POD and Nanobiohybrid were checked at 70 and 90 °C. Aliquots were withdrawn at regular time intervals for assaying the residual enzyme activities. The storage stabilities of the free 3,4-POD and Nanobiohybrid were determined at 4 and 25 °C for 1 month. The residual activities of the free 3,4-POD and Nanobiohybrid were measured at regular day intervals.

The recycling of Nanobiohybrid was performed according to the following methods. A 300 μL of 1.0 mg/ml Nanobiohybrid was mixed into 100 μM of 3,4-DHBA, and the activities were assayed as outlined in Sect. 6.3.2. The Nanobiohybrid was then recovered by centrifugation (14,000 rpm, 10 min), and washed 5 times with fresh CHES buffer (50 mM, pH 10.0) in order to remove the residual 3,4-DHBA and its oxidized products. A total of ten cycles was performed, and the results were expressed as an average of three replicates.

6.3.4 Removal of 3,4-DHBA

Batch experiments were performed in 10.0 mL screw tabs sealable glass reaction bottles containing 2.0 mg of F-MWCNTs as control, free 3,4-POD and Nanobiohybrid in CHES (50 mM, pH 9.0). All of these were treated with 5.0 mM of 3,4 DHBA solutions. The reaction mixtures were stirred at 200 rpm at 25 ± 1 °C. Aliquots were withdrawn at regular time intervals for measuring the concentrations of 3,4-DHBA at 290 nm [28, 29]. The amount of 3,4-DHBA biodegraded by the immobilized 3,4-POD in the Nanobiohybrid was calculated using Eq. (6.3).

$$Q_T = Q_I - Q_F - Q_A \quad (6.3)$$

where Q_T is the amount of 3,4-DHBA (mM) degraded by the Nanobiohybrid; Q_I is the initial 3,4-DHBA concentration (mM) in the solution; Q_F is the amount of 3,4-DHBA (mM) retained in the solution; and Q_A is the 3,4-DHBA concentration (mM) adsorbed into the F-MWCNTs. All treatments were replicated 5 times, and the average values were obtained.

6.3.5 *Data Analysis*

Experimental data were statistically analyzed using Origin (version 7.0) and Sigma plots (version 10.0). Data were weighted significant when $p < 0.05$. Data were fitted using suitable nonlinear regression analysis methods. Experimental values of enzyme kinetics were evaluated using GraphPad Prism 6.0 for windows available at (http://www.graphpad.com/), and replicates were performed multiple times to evaluate inadequate model. In order to determine actual biodegradation rate of Nanobiohybrid, we used first-order hopane-normalized model [26] as shown in Eq. (6.4). It helps us to measure the rate constant (k), half life ($t_{1/2}$), and the RE of 3,4-DHBA as shown in Eqs. (6.4), (6.5), and (6.6), respectively

$$C_t = C_0 \exp(-kt), \quad (6.4)$$

where C_0 and C_t, are the 3,4-DHBA concentrations at the time 0 and t (h), respectively. Half-life is in the time (h) that is computed as follows:

$$t_{1/2} = \ln(2)/k \quad (6.5)$$

$$RE_t = (C_0 - C_t)/C_0 x\ 100 \quad (6.6)$$

6.4 Results

6.4.1 *Effects of pH and Temperature on Free 3,4-POD and Nanobiohybrid Activities*

Figure 6.4 shows the pH (a) and temperature (b) behaviors of free 3,4-POD and Nanobiohybrid relative actvities. It describes the activity profiles of the free 3,4-POD and Nanobiohybrid were followed similar trends up to pH 8.0 and temperature 50 °C, after which the activities were varied. The optimum activity was shifted from pH 9.0 to 10.0 and temperature from 55 to 60 °C for the free 3,4-POD and Nanobiohybrid, respectively, that were similar to earlier observations [11]. The

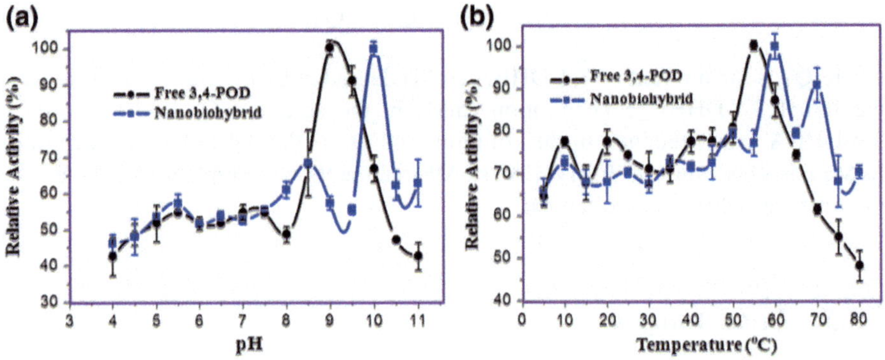

Fig. 6.4 Effects of **a** pH and **b** temperature on 3,4-POD activities

Nanobiohybrid showed significantly higher relative activities of 63 and 70% than the free 3,4-POD which were 42 and 48% (p < 0.05) at pH 11.0 and 80 °C, respectively.

6.4.2 *Kinetic Analyses of Free 3,4-POD and Nanobiohybrid*

The relative activity, and all the essential kinetic parameters measured from the Michaelis–Menten plots as shown in Figs. 6.5 and 6.6 of the free 3,4-POD and Nanobiohybrid, respectively, are also revealed in Table 6.1. The relative activity retained by the Nanobiohybrid was 93% to its free 3,4-POD. As a common rule, the Nanobiohybrid showed 2.5 times higher K_m than the free 3,4-POD. But Nanobiohybrid had a higher turnover number and significantly increased velocity (V_{max}) than the free 3,4-POD (p < 0.05), suggesting appropriate support of MWCNTs [20]. Meanwhile, Nanobiohybrid showed overall catalytic efficiency (K_{cat}/K_m) almost 50% to its free 3,4-POD, indicating the support endows suitable catalytic efficiency of Nanobiohybrid for degrading 3,4-DHBA.

Fig. 6.5 Michaelis–Menten plot of the free 3,4-POD

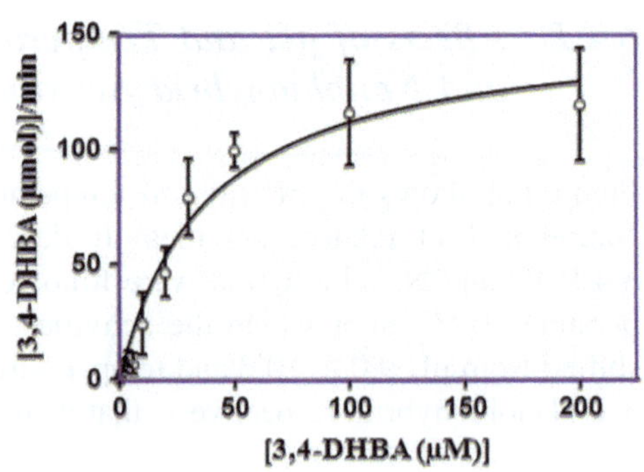

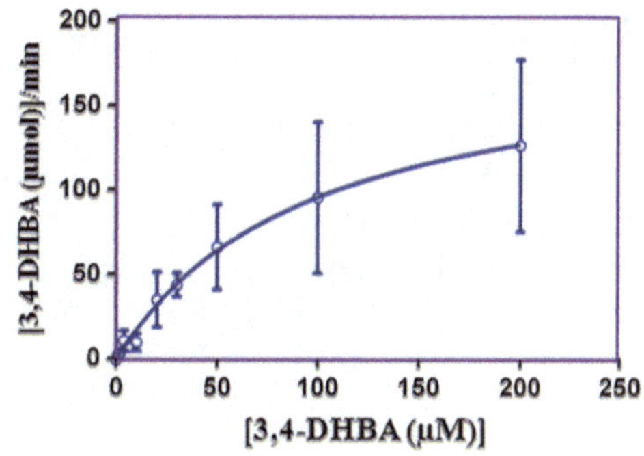

Fig. 6.6 Michaelis–Menten plot of the Nanobiohybrid

Table 6.1 Relative activity and kinetic parameters for free 3,4-POD and Nanobiohybrid

3,4-POD	Relative Activity[a](%)	V_{max} (µmole/min)	K_m (µM)	K_{cat} (min^{-1})	K_{cat}/K_m ($min^{-1}.\mu M^{-1}$)
Free	–	153 ± 24	37 ± 15	210 ± 33	5.61 ± 3
Nanobiohybrid	93	188 ± 69	95 ± 71	258 ± 94	2.71 ± 3

[a]The relative activity was determined by (Specific activity of Nanobiohybrid/Specific activity of free 3,4 POD) x 100

6.4.3 Stability Studies of Free 3,4-POD and Nanobiohybrid

Figure 6.7 shows the effects of temperatures (a,b) and storage (c) on the residual activities of free 3,4-POD and Nanobiohybrid. According to Fig. 6.7a, b, Nanobiohybrid was significantly more stable than the free 3,4-POD ($p < 0.05$) at both 70 (Fig. 6.7a) and 90 °C (Fig. 6.7b). The free 3,4-POD lost its activities 43 and 82%, while Nanobiohybrid was 38 and 66% after 300 and 180 min incubations at 70 and 90 °C, respectively. As shown in Fig. 6.7c, the activities of the free 3,4-POD were abated significantly than the Nanobiohybrid at 4 and 25 °C. Nanobiohybrid retained its 56% activity which was 41 and 39% for the free 3,4-POD at 4 and 25 °C after 30-day storages ($p < 0.05$), respectively. Figure 6.7d represents the recycling of Nanobiohybrid catalyst. It shows > 97% (mean) of the Nanobiohybrid activity was retained up to five batch uses. After ten cycle uses, the activity was retained around 60%, suggesting greater operational stability of the developed hybrid catalyst.

6.4.4 Degradation Kinetics of 3,4-DHBA

Figure 6.8 shows the removal percentages of 3,4-DHBA in a 6 h batch experiment. The data were fitted to first-order reaction, and the important kinetic parameters with the RE are listed in Table 6.2. It shows the slower removal rate of

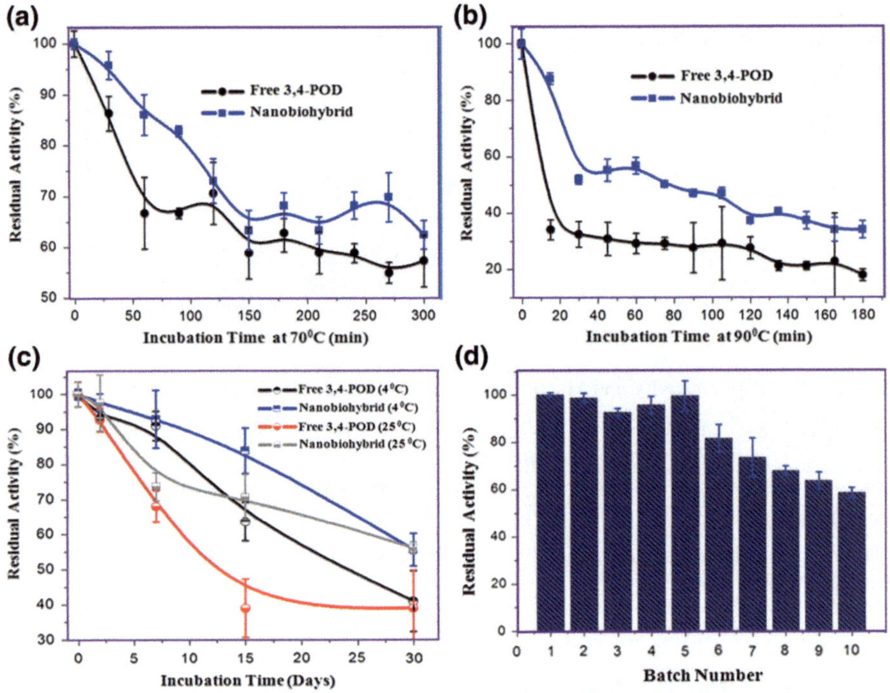

Fig. 6.7 Thermostabilities at **a** 70 and **b** 90 °C, storage stabilities of the free and Nanobiohybrid **c**, and reusability of Nanobiohybrid catalyst **d**

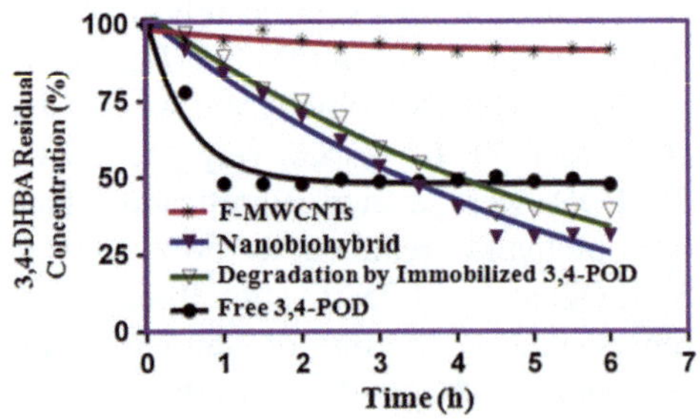

Fig. 6.8 Removal kinetics of 3,4-DHBA by free and Nanobiohybrid

Nanobiohybrid ($t_{1/2}$: 3.81 h) than the free enzyme ($t_{1/2}$: 0.37 h). However, the effects were found temporary, since higher RE (71%) was observed for Nanobiohybrid compared with 54% of the free 3,4-POD with increasing time. Only 7% RE of F-MWCNTs was observed, which might attribute due to its adsorption behavior. It represented negligible removal of 3,4-DHBA from wastewater by the F-MWCNT alone. This further supported the desirable biodegradation effects of Nanobiohybrid toward the 3,4-DHBA removal (64%).

Table 6.2 Reaction Kinetic Parameters and Removal Efficiency (RE) of the free and Nanobiohybrid toward 3,4-DHBA

Sample	K^{-1} (h)	$t_{1/2}$ (h)	RE_6 (%)
F-MWCNTs Alone	0.40	1.71	7
Nanobiohybrid	0.18	3.81	71
Degradation by Immobilized 3,4-POD	0.16	4.26	64
Free 3,4-POD	1.85	0.37	54

6.5 Discussion

The success of a Nanobiohybrid catalyst for practical applications usually relies on certain parameters such as pH and temperature. As shown in Fig. 6.4, Nanobiohybrid was potentially active in higher alkaline and temperature ranges. These improvements of the activities amending higher stability of the Nanobiohybrid than the free 3,4-POD. The effects might be due to the increase of enzyme's rigidity [12], MWCNT properties and the microenvironmental changes that could affect the conversion of substrate to product [32]. Higher Nanobiohybrid activity at more alkaline pH was probably because of the slower diffusion of cis, cis-muconic acid from the active sites into the external solvenst phases [32]. In addition, F-MWCNTs could have more free acidic groups that shifted alkali microenvironment for higher Nanobiohybrid activities.

Table 6.1 shows that the Nanobiohybrid could keep 93% of relative activity to its free 3,4-POD, which was highest than the activities of immobilized 3,4-POD on porous glass (70 and 90%) [28, 29] and agarose (42 and 68%) [28, 29]. This high activity might be due to the uses of appropriate support and immobilization method, optimized catalyst reaction conditions and 3,4-POD sources. However, as a general trend, we observed a decrease in affinity of binding between 3,4-DHBA and Nanobiohybrid compared with free 3,4-POD because of increasing K_m which is still lower than the K_m of the 3,4 POD immobilized on agarose (15–20 times greater than the free enzyme), suggesting less mass transfer limitations and conformational changes [25] of 3,4-POD. The result is consistent with earlier CD studies as described in Chap. 5, Sect. 5.4.2 that showed 44% relative structural change of Nanobiohybrid to its free 3,4-POD configurations. This might be the possible way to decrease about 50% of Nanobiohybrid overall catalytic efficiency (K_{cat}/K_m) (Table 6.1). Cang-Rong and Pastorin [4] hypothesized that an immense structural retention of a Nanobiohybrid could lead to a greater retention of catalytic rate and efficiency to its free enzyme [4].

The mattering much to immobilize an enzyme onto a support is to intensify its shelf life at extreme conditions, which affects its viability for industrial applications. Since peaked activities of the free 3,4-POD and Nanobiohybrid were observed at high temperature ranges (Fig. 6.4b), which allowed us to check its thermostabilities at higher temperatures such as 70 and 90 °C as shown in Fig. 6.7 (a, b). It shows higher thermostability of the Nanobiohybrid compared with free 3,4-POD and earlier results [32]. Zaborsky and Ogletree [32] found no significant changes of enzyme activities of the free and immobilized 3,4-POD onto agarose. The free and immobilized 3,4-POD were inactivated after 60 and 75 min at 60 °C only [32]. Therefore, the developed Nanobiohybrid catalyst in this study could play better role in purifying hot industrial effluents. The thermostability of Nanobiohybrid might be due to the support mediated conformational changes of 3,4-POD, nature of support and reduced molecular mobility [27]. Moreover, the Nanobiohybrid showed higher residual activities than the free 3,4-POD on storage for a month at 4 and 25 °C (Fig. 6.7c), attributing better catalytic stabilities. Guzik et al. [11] observed less storage stabilities, since immobilized 3,4-POD onto alginate and agarose was capable to retain only 30 and 10% of residual activities after 21 and 28 days, respectively. It clarifies that the our developed Nanobiohybrid was significantly resistant to inhibition of its activity changes by longer incubation periods than the free 3,4-POD and other immobilization conditions. The chemical bondings between F-MWCNTs and 3,4-POD could prevent structural denaturation of the enzyme. Similar hypothesis had been reported for laccase enzyme [6]. The enhancement of these overall storage stabilities of the developed system would give a sufficient time for transportation of Nanobiohybrid to remote uses in water purification reactor.

Appropriate physical properties of MWCNTs [7] had allowed us to separate Nanobiohybrid from the reaction mixture. It would decrease the production cost of 3,4-POD in catalytic applications and subsequently increase the feasibility of the developed system to use industrially. Higher recycling of Nanobiohybrid was observed as shown in Fig. 6.7d. Although no reusability data of immobilized 3,4-POD was presented in previous studies [11, 23, 28, 29, 32], similar data can be comparable to other enzymes immobilized on different supports [10, 30]. Therefore, the developed Nanobiohybrid could compensate the production costs of 3,4-POD in decontaminating 3,4-DHBA mixed water.

The cytotoxicity of 3,4-DHBA to human cell lines had been suggested to occur at 5.0 mM ($p \leq 0.01$), whereas nontoxic level was from 1.0 to 2.5 mM [1]. Hence, it is not unreasonable to investigate the removal of 5.0 mM 3,4-DHBA by our developed and optimized Nanobiohybrid as shown in Fig. 6.8 and Table 6.2. Although the degradation efficiency of free 3,4-POD was higher ($t_{1/2}$: 0.37 h) than the Nanobiohybrid ($t_{1/2}$: 3.81 h), the higher RE of Nanobiohybrid (71%) could compensate it over time. The phenomenon is consistent with earlier enzyme reaction kinetics described in Sect. 6.4.2. This is because of 3,4-DHBA diffusion limit into the 3,4-POD active sites and low molecular flexibility upon immobilization onto support material [27]. Another hypothesis could be the *cis,cis*-muconic acid that was formed from the 3,4-DHBA breakdown may create steric blockage once its release into the external milieu of enzyme's active site was disrupted. It may promote the temporarily hiatus of further binding of substrates to the active sites, which leads to the decrease in breakdown rate of Nanobiohybrid, but could consistently overcome the obstruction with increasing time. Similar observations can be comparable for other enzymes mediated biodegradation of water pollutants [15, 31].

6.6 Conclusions

The present study reported the uses of Nanobiohybrid for the effective degradation of 3,4-DHBA in water. Compared with free 3,4-POD, Nanobiohybrid showed greater stabilities in higher alkaline pH and temperature zones. The free 3,4-POD lost its residual activity of 82%, while Nanobiohybrid was 66% after 180 min incubations at 90 °C. Moreover, Nanobiohybrid could retain 93% and about 50% of its relative activity and overall catalytic efficiency to the free 3,4-POD, respectively. Higher storage stability of the Nanobiohybrid was observed, since it maintained >55% of residual activity compared with free 3,4-POD, which was almost 40% after 30 days of storage at both 4 and 25 °C. Recrudescent Nanobiohybrid could keep >60% of residual activity after ten operational cycles used, endowing to decrease the production costs of 3,4-POD for long-term uses. More than 70% of 3,4-DHBA removed by the Nanobiohybrid in less than 4 h treatment, suggesting a reduced time protocol. Therefore, with these overall results analyses we can conclude that the developed Nanobiohybrid here could act as an efficient novel decontamination platform for mineralizing 3,4-DHBA in water.

References and Future Readings

1. Babich, H., Sedletcaia, A., Kenigsberg, B.: In vitro cytotoxicity of protocatechuic acid to cultured human cells from oral tissue: involvement in oxidative stress. Pharmacol. Toxicol. **91** (5), 245–253 (2002)
2. Benitez, F.J., Beltran-Heredia, J., Acero, J.L., Gonzalez, T.: Degradation of protocatechuic acid by two advanced oxidation processes: Ozone/UV radiation and H2O2UV radiation. Water Res. **30**(7), 1597–1604 (1996)
3. Buchan, A., Collier, L.S., Neidle, E.L., Moran, M.A.: Key aromatic-ring-cleaving enzyme, protocatechuate 3, 4-dioxygenase, in the ecologically important marineRoseobacter lineage. Appl. Environ. Microbiol. **66**(11), 4662–4672 (2000)
4. Cang-Rong, J.T., Pastorin, G.: The influence of carbon nanotubes on enzyme activity and structure: investigation of different immobilization procedures through enzyme kinetics and circular dichroism studies. Nanotechnology **20**(25), 255102 (2009)
5. Cloete, T.E.: Nanotechnology in Water Treatment Applications. Horizon Scientific Press (2010)
6. Dai, Y., Yin, L., Niu, J.: Laccase-carrying electrospun fibrous membranes for adsorption and degradation of PAHs in shoal soils. Environ. Sci. Technol. **45**(24), 10611–10618 (2011)
7. Das, R., Abd Hamid, S.B., Ali, M.E., Ismail, A.F., Annuar, M.S.M., Ramakrishna, S.: Multifunctional carbon nanotubes in water treatment: the present, past and future. Desalination **354**, 160–179 (2014). doi:10.1016/j.desal.2014.09.032
8. Das, R., Hamid, S.B.A., Ali, M.E., Ismail, A.F., Annuar, M., Ramakrishna, S.: Multifunctional carbon nanotubes in water treatment: the present, past and future. Desalination **354**, 160–179 (2014)
9. Gernjak, W., Krutzler, T., Glaser, A., Malato, S., Caceres, J., Bauer, R., Fernández-Alba, A.: Photo-Fenton treatment of water containing natural phenolic pollutants. Chemosphere **50**(1), 71–78 (2003)
10. Goh, W.J., Makam, V.S., Hu, J., Kang, L., Zheng, M., Yoong, S.L., Udalagama, C.N., Pastorin, G.: Iron oxide filled magnetic carbon nanotube-enzyme conjugates for recycling of

amyloglucosidase: toward useful applications in biofuel production process. Langmuir **28** (49), 16864–16873 (2012)
11. Guzik, U, Hupert-Kocurek, K, Krysiak, M, Wojcieszyńska, D.: Degradation potential of protocatechuate 3, 4-dioxygenase from crude extract of stenotrophomonas maltophilia strain kb2 immobilized in calcium alginate hydrogels and on glyoxyl agarose. BioMed Res. Int. (2014)
12. Iyer, P.V., Ananthanarayan, L.: Enzyme stability and stabilization—aqueous and non-aqueous environment. Process Biochem. **43**(10), 1019–1032 (2008)
13. Johnson, K.A., Goody, R.S.: The original Michaelis constant: translation of the 1913 Michaelis-Menten paper. Biochemistry **50**(39), 8264–8269 (2011). doi:10.1021/bi201284u
14. Khin, M.M., Nair, A.S., Babu, V.J., Murugan, R., Ramakrishna, S.: A review on nanomaterials for environmental remediation. Energy Environ. Sci. **5**(8), 8075–8109 (2012)
15. Lante, A., Crapisi, A., Krastanov, A., Spettoli, P.: Biodegradation of phenols by laccase immobilised in a membrane reactor. Process Biochem. **36**(1), 51–58 (2000)
16. Mechrez, G., Krepker, M.A., Harel, Y., Lellouche, J.-P., Segal, E.: Biocatalytic carbon nanotube paper: a 'one-pot'route for fabrication of enzyme-immobilized membranes for organophosphate bioremediation. J. Mater. Chem. B **2**(7), 915–922 (2014)
17. Nakamura, Y., Torikai, K., Ohigashi, H.: Toxic dose of a simple phenolic antioxidant, protocatechuic acid, attenuates the glutathione level in ICR mouse liver and kidney. J. Agric. Food Chem. **49**(11), 5674–5678 (2001)
18. Nakamura, Y., Torikai, K., Ohto, Y., Murakami, A., Tanaka, T., Ohigashi, H.: A simple phenolic antioxidant protocatechuic acid enhances tumor promotion and oxidative stress in female ICR mouse skin: dose-and timing-dependent enhancement and involvement of bioactivation by tyrosinase. Carcinogenesis **21**(10), 1899–1907 (2000)
19. Neun, D.J., Penn, A., Snyder, C.A.: Evidence for strain-specific differences in benzene toxicity as a function of host target cell susceptibility. Arch. Toxicol. **66**(1), 11–17 (1992)
20. Ren, L., Zhong, W.: Oxidation reactions mediated by single-walled carbon nanotubes in aqueous solution. Environ. Sci. Technol. **44**(18), 6954–6958 (2010). doi:10.1021/es101821m
21. Rivas, F.J., Frades, J., Alonso, M.A., Montoya, C., Monteagudo, J.M.: Fenton's oxidation of food processing wastewater components. kinetic modeling of protocatechuic acid degradation. J. Agric. Food Chem. **53**(26), 10097–10104 (2005). doi:10.1021/jf0512712
22. Sarma, J., Mahiuddin, S.: Specific ion effect on the point of zero charge of α-alumina and on the adsorption of 3,4-dihydroxybenzoic acid onto α-alumina surface. Colloids Surf., A **457**, 419–424 (2014). doi:10.1016/j.colsurfa.2014.06.014
23. Silva, ASd, Jacques, R.J.S., Andreazza, R., Bento, F.M., Camargo, FAdO: The effects of trace elements, cations, and environmental conditions on protocatechuate 3, 4-dioxygenase activity. Sci. Agricola **70**(2), 68–73 (2013)
24. Stanier, R., Ingraham, J.: Protocatechuic acid oxidase. J. Biol. Chem. **210**(2), 799–808 (1954)
25. Subrizi, F., Crucianelli, M., Grossi, V., Passacantando, M., Pesci, L., Saladino, R.: Carbon nanotubes as activating tyrosinase supports for the selective synthesis of catechols. ACS Catal. **4**(3), 810–822 (2014)
26. Venosa, A., Suidan, M., King, D., Wrenn, B.: Use of hopane as a conservative biomarker for monitoring the bioremediation effectiveness of crude oil contaminating a sandy beach. J. Ind. Microbiol. Biotechnol. **18**(2–3), 131–139 (1997)
27. Wang, F., Guo, C., Liu, H.Z., Liu, C.Z.: Immobilization of Pycnoporus sanguineus laccase by metal affinity adsorption on magnetic chelator particles. J. Chem. Technol. Biotechnol. **83**(1), 97–104 (2008)
28. Wojtaś-Wasilewska, M., Luterek, J., Rogalski, J.: Immobilization of protocatechuate 3, 4-dioxygenase from *Pleurotus ostreatus* on activated porous glass beads. Phytochemistry **27** (9), 2731–2733 (1988)
29. Wojtaś-Wasilewska, M., Luterek, J., Leonowicz, A., Dawidowicz, A.: Dearomatization of lignin derivatives by fungal protocatechuate 3, 4-dioxygenase immobilized on porosity glass. Biotechnol. Bioeng. **32**(4), 507–511 (1988)

30. Xu, R., Chi, C., Li, F., Zhang, B.: Laccase-Polyacrylonitrile Nanofibrous Membrane: Highly Immobilized, Stable, Reusable, and Efficacious for 2, 4, 6-Trichlorophenol Removal. ACS Appl. Mater. Interfaces **5**(23), 12554–12560 (2013)
31. Xu, R., Zhou, Q., Li, F., Zhang, B.: Laccase immobilization on chitosan/poly (vinyl alcohol) composite nanofibrous membranes for 2, 4-dichlorophenol removal. Chem. Eng. J. **222**, 321–329 (2013)
32. Zaborsky, O.R., Ogletree, J.: Immobilization of protocatechuate 3, 4-dioxygenase with activated agarose. Biochim. Biophys. Acta (BBA)-Enzymol. **289**(1), 68–76 (1972)

MIX
Papier aus verantwortungsvollen Quellen
Paper from responsible sources
FSC® C105338

If you have any concerns about our products,
you can contact us on
ProductSafety@springernature.com

In case Publisher is established outside the EU,
the EU authorized representative is:
Springer Nature Customer Service Center GmbH
Europaplatz 3, 69115 Heidelberg, Germany

Printed by Libri Plureos GmbH
in Hamburg, Germany